The Open University

Technology Foundation Course Unit 2

The week number during which this unit should be studied is not necessarily the same as the unit number. Please consult your wall-chart study guide to the Technology Foundation Course to find the place of this unit in the course.

THE HUMAN COMPONENT

Prepared by John Beishon and Peter Zorkoczy for the Technology Foundation Course Team

THE OPEN UNIVERSITY PRESS

The Technology Foundation Course Team

G. S. Holister (*Chairman and General Editor*)
K. Attenborough (*Engineering Mechanics*)
R. J. Beishon (*Systems*)
D. A. Blackburn (*Materials Science*)
J. K. Cannell (*Engineering Mechanics*)
A. Clow (*BBC*)
G. P. Copp (*Assistant Editor*)
D. G. Crabbe (*Course Assistant*)
C. L. Crickmay (*Design*)
N. G. Cross (*Design*)
E. S. L. Goldwyn (*BBC*)
J. G. Hargrave (*Electronics*)
R. D. Harrison (*Educational Technology*)
M. J. L. Hussey (*Engineering Mechanics*)
A. B. Jolly (*BBC*)
J. C. Jones (*Design*)
L. M. Jones (*Systems*)
R. D. R. Kyd (*Editor*)
J. McCloy (*BBC*)
R. McCormick (*Educational Technology*)
D. Nelson (*BBC*)
C. W. A. Newey (*Materials Science*)
S. Nicholson (*Design*)
G. Peters (*Systems*)
A. Porteous (*Engineering Mechanics*)
C. Robinson (*BBC*)
R. Roy (*Design*)
J. J. Sparkes (*Electronics*)
R. Thomas (*Economics*)
G. H. Weaver (*Materials Science*)
P. I. Zorkoczy (*Electronics*)
and the late Professor R. K. Ham (*Materials Science*)

The Open University Press
Walton Hall, Milton Keynes

First published 1971. Reprinted 1973 (with corrections)

Designed by the Media Development Group of the Open University

Printed in Great Britain by
MARTIN CADBURY, A SPECIALIZED DIVISION OF SANTYPE INTERNATIONAL, WORCESTER AND LONDON

SBN 335 02501 3

Open University courses provide a method of study for independent learners through an integrated teaching system, including text material, radio and television programmes and short residential courses. This text forms part of a series that makes up the correspondence element of the Technology Foundation Course.

The Open University's courses represent a new system of university-level education. Much of the teaching material is still in a developmental stage. Courses and course materials are, therefore, kept continually under revision. It is intended to issue regular up-dating notes as and when the need arises, and new editions will be brought out when necessary.

For general availability of supporting material referred to in this book, please write to the Director of Marketing, The Open University, P.O. Box 81, Milton Keynes, MK7 6AT.

Further information on Open University courses may be obtained from the Admissions Office, The Open University, P.O. Box 48, Milton Keynes, MK7 6AB.

1.2

Objectives

The primary aim of this unit is to draw attention to the characteristics of human beings which the technologist must take into account when designing a system to match its human operators and to meet the needs of its users. Attention is concentrated particularly on the way in which we respond to and interpret signals from the outside world and especially on the characteristics of the eye and the ear. These are of importance in many branches of technology such as telecommunications, instrumentation, traffic control and illumination engineering.

After you have read the unit you should be able to:

(1) Pick out good and bad examples of design in the devices used in the home and explain why you think they are good or bad.

(2) Describe the salient features of our sensory and motor organs.

(3) Describe the main features of the eye, and explain in general terms how it works and state the limits of perception of the normal human eye, and how these vary with intensity of illumination and wavelength.

(4) Discuss ways in which we perceive spatial relationships and give examples of some of the ways in which a misleading impression may be conveyed.

(5) Give examples of the ways in which relative judgements are more sensitive than absolute ones.

(6) Describe the main features of the ear, explain in general terms how the ear works and state the limits of perception of the normal human ear and how these vary with pitch.

(7) Describe how we locate the source of a sound.

(8) Discuss the way we make decisions and the nature of skill.

(9) Describe and discuss applications of the ideas introduced in this unit in familiar situations such as driving a car, typing a letter or playing a musical instrument.

What you have to do

There are five simple exercises in Section 3, one in Section 6 and two in Section 7.

In Sections 3.1, 3.4 and 6 there are simple experiments for which you will require the help of another person.

In addition in Sections 3, 6 and 7 there are four examples for you to work out, with answers given on p. 44.

No self-assessment questions have been set; for revision, check yourself against the objectives and go over the assignment questions.

Contents

Introduction

The title of our foundation course is 'The man-made world' and there is surely no need to have to draw anyone's attention to the impact that man-made things have made and still make on us. The enormous growth and success of man-made objects over the last 200 years has produced more changes in our way of living than probably took place in the previous 2 000, or even 20 000. In our modern industrialized culture we are surrounded by man-made tools, machines and implements and we would find it difficult to survive now without them. In the past, the development of tools and implements took place only relatively slowly and there was ample time for each object to evolve into a form suitable for human use. The plough, the scythe, simple weapons, these were all modified and improved slowly so that even now some of these tools are almost unchanged in shape from those we used a thousand years ago.

The industrial revolution and the development of technology have changed this situation in two ways. Firstly, the pace of change has meant that tools, implements and equipment have altered their form much more rapidly, faster in fact than the rate at which the adaptation of their shape and performance to suit human needs could be accommodated. Secondly, man has produced devices and machines whose performance goes far beyond the range of his own normal sensory experience. Until relatively recently, man had never seen an object smaller than 0·1 mm and, correspondingly, he could not even see ordinary sized objects if they were very far away. He would rarely have experienced any object moving much faster than, say, a bird flying or a stone falling at perhaps 30–50 mph. Certainly he was rarely called upon to interact with anything which moved or changed position very quickly. Now, even with simple machines, a very wide range of speeds and movement effects have become part of the *direct* experience of human beings.

These two effects, the rapidity of change and the extension of the range of effects, have had a number of results. The pace of change meant that people were forced to adapt their behaviour to that of the machine and they had to try and cope with objects that were shaped by the overriding considerations of engineering needs and economic factors. The modern lathe today is essentially very like the original machine and it really needs a 4 ft dwarf with an 8 ft arm-spread and a swan-neck to operate it effectively. It is probably not much of an exaggeration to say that even now very few objects in our lives are really designed to suit our particular human needs and limitations. Car seats are still giving us back aches, lights and visual displays causing glare and eye strain, and noise distracting and disturbing our concentration and enjoyment of life. A recent report* reveals that

> Thirty-five million work days are lost each year because of arthritis, and the cost is estimated at around £170 million

and it suggests that

> because of their job, some workers are more liable to get arthritis than others. Studies have shown that in a year coal miners lost 266 work days for every 100 workers because of rheumatism. With unskilled labourers, the figure was

---

* *The Guardian*, 19 April 1971.

203 lost days, but with clerks only 81. Terms like housemaid's knee, policeman's heel, weaver's bottom and telegraphist's wrist all had occupations incorporated in them.

Of course there are many complex reasons why so much around us is inconvenient or dangerous or ill-suited to our different shapes and sizes, and later units in this course will explore these in greater depth. But in a surprisingly large number of cases there appears to be no good reason why the artificial environment does not suit us, other than the simple fact that we neglected to take ourselves into account when we designed it. (It might be more accurate to say that the equipment or environment *evolved* rather than that we ever *designed* it.)

One reason why this situation has persisted for a very long time is that we are such adaptable animals. Our extremely sensitive and adjustable hand–finger system, for example, enables us to pick up, hold and manipulate an enormous variety of objects. This has meant that we have paid little attention to the design of hand grips until recently. Now some pistol grip drills have finger indentations for easier and firmer grasping. Our bodies are almost infinitely adjustable so we often find ourselves adopting very uncomfortable and strained positions to operate machines simply because we *can* get into these positions; it is not until we are forced to adopt them for hours or weeks on end that we begin to realize the discomforts and longer-term deleterious effects. If we could not have coped in the first place, the machine would have *had* to have been redesigned.

The man-made objects around us are designed for a purpose and very few of them can operate completely independently of us. Even with fully automatic equipment there is always a need for maintenance or repair work and in the end a man always has to grapple with the object itself in some way. With most of our tools, implements and equipment, we use the device as an extension or transformation of our muscle power.

We 'relegate' ourselves to being guiders, pilots or *controllers* of the machine, so that we determine what in particular it will do or achieve from moment to moment. The motor car transports us about and we put very little muscular effort into it these days. The brakes and steering are often power-assisted, the gear changes automatic; we can drive almost literally with our finger tips and control the motion and direction of up to a ton of machine plus ourselves with ease. More and more we are using ourselves as the *controlling* or *directing* part of a complex system designed for some purpose where the physical effort or power is provided by other energy sources.

Even this relatively new function of controller is being taken over by machine devices to some extent. Some writers refer now to a *second* industrial revolution; the first came, of course, 200 years ago when we started to replace our muscle power with mechanical, electrical and chemical energy; this new second one is starting now as we begin a similar process of replacing our mental effort by machines or by using machines themselves to control and direct effort. A fully-laden aircraft can land completely automatically without a pilot having to see the runway or control the aircraft at all. Computing and control devices do what the pilot would normally do and in some cases do it more safely and more effectively than the pilot can, such as making a landing in fog. However, this new revolution is only just beginning and there is still much controversy about where it is heading and how far man can, or will, allow himself to be replaced in this way. We are still faced, and probably will be for many decades yet, with the problem of fitting man and his man-made objects together in mutually effective and satisfactory ways. For this reason we shall be looking more closely at both man and machine in this course, and the remainder of this unit will be devoted to a study of the *human component* in these man–machine combinations.

Section 2

Man and machine

For convenience at this stage let us consider only a relatively simple system of one man and one machine: let's take a typical car and its driver. We will have to ignore many interesting aspects of the situation in this unit because they would take us beyond our brief—we shall not, for example, consider *why* the man is using the car, or whether he is in a hurry, or angry, or timid, or even if 'he' is a woman! We shall confine our attention to the *cognitive* aspects of the situation, that is the rational, thinking, voluntary part of man's behaviour, the part that we use for doing skilled tasks, for learning new things both physical and mental. Obviously, we cannot just ignore the motivational and emotional aspects of human behaviour but in this unit we are more interested in man's use of his thinking abilities to control, direct and interact with man-made devices. The basic picture we shall be working with is shown in Fig. 1.

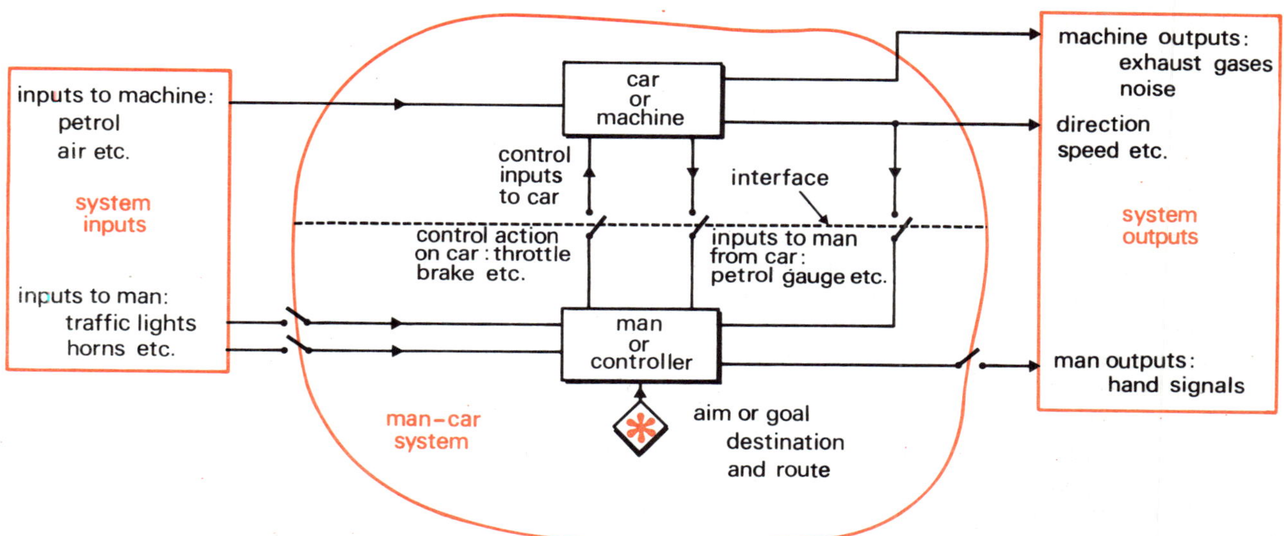

Fig. 1   Simplified model of man–machine (car) system. The inputs to the system are divided into those which go direct to the machine and those going to the man. The outputs are shown as three separate types—from man, machine or from the system as a whole. Note that the man's interactions, both with the machine and the system environment, are carried out via switches which indicate that the human operator acts in a discrete way and has to switch attention from one channel to another.

This shows a model of the situation, a simplified model, but one which illustrates the main features we shall be considering. We have used the example of a 'man driving car' to give some idea of the *inputs* and *outputs* in the system. This is a general picture, however, and it can be applied to a wide range of man–machine systems although in each system the actual components or functions may differ.

Our man is depicted as a *controlling box* with inputs from the system environment in general, for example sensing of traffic conditions, and also inputs directly from the machine in question, petrol level for example. We also have outputs from the man which can take the form of direct muscular actions on the car, pressing the brake pedal for example, but which may also be information outputs from the system, such as signalling a right turn. The inputs and outputs between man and car go across a machine–man *interface*, and the system boundary is regarded as an invisible envelope between the man/car and the outside world. This envelope is not identical with the bodywork of the car and it can extend into the outside world as, for example, when our driver sees and responds to a red traffic light some hundred yards away. Just where you draw the system boundary in this case depends partly on what you are trying to achieve and partly on convenience.

You will see that the inputs to, and outputs from, the man are not direct connections but go via 'switches'. These are meant to show that man does not respond continuously to the car and the system environment but that he receives or puts out information or actions in discrete amounts. This is easily illustrated by the way we look, for example, at the speedometer, then at the road in front, then at the rear-view mirror. We cannot watch any of these continuously and we have to 'switch' our attention from one to another as appropriate. We can do a limited number of things at the same time, such as watch a car in front and change gear, but in general we can only attend to one thing at a time.

When studies are made in detail of human actions it appears that virtually all controlled human action takes place in tiny separate motions. These motions are sufficiently small for the overall actions to appear as one smoothly flowing movement but there are very good reasons for regarding the outputs of the human being as discrete rather than continuous actions. This is not true of certain '*ballistic*' motions of the limbs as when you allow your limbs to move more freely under gravity or after a muscle impulse.

It is important to note here that what crosses the interface between man and the outside world is mainly information. It is true that this is carried by an energy form such as light or sound, but the carrier is largely unimportant, it is the *nature* of the information which is important. This is more true of inputs than outputs because we are more used to acting physically on the environment, but occasionally we receive a physical input such as a punch on the nose. Note that this also conveys information. The machine or car in our diagram (Fig. 1) also has inputs and outputs. Some of these are obvious enough—petrol and air as inputs for example, and exhaust gases as outputs—but, as with defining a system, we tend to identify as system outputs those aspects of the behaviour of the machine which interest us. In the case of a car we are more interested in its velocity than in its heat output. Notice that we have classified some of the inputs to the machine as *control inputs*, such as throttle or brake controls, because these enable the driver to exercise control over the system output he is interested in.

Section 3

Information processing

Our example showed one very familiar man–machine system. We shall now concentrate on man, the human component. As we have already mentioned, man is increasingly becoming a non-manual controller of machines and processes and, instead of handling objects and using physical force, he is handling *information*. In the current jargon, man is regarded as an *information processing device* and we are interested in studying how he obtains information, how he decides what to do with it, and how he takes action on the basis of the decision. A simple but useful way of looking at man as a *system component*, for these purposes, is to regard him as a device with input and output (Fig. 2).

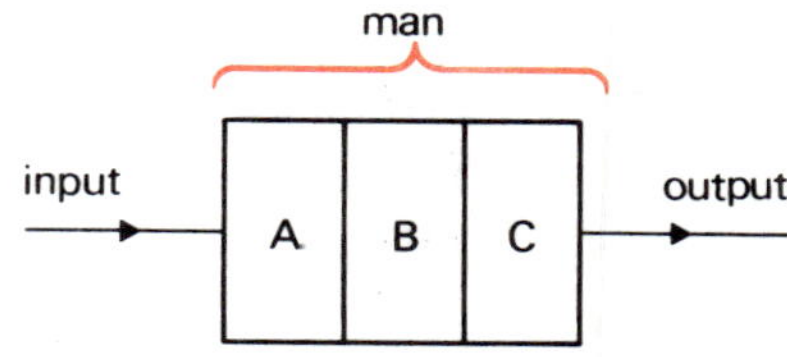

Fig. 2.

sensory and perceptual processes

decision processes

motor processes

The parts labelled A, B and C refer very loosely to internal processes which we know must take place within the human nervous system. The A part takes in information from the senses and organizes it in some meaningful way; we call these *sensory and perceptual processes*. The B part uses the incoming organized *perceptions* and makes decisions about what to do (or not to do); we call this the *decision process*. The C part translates the decisions into *actions*, often movements or coordinated motor actions of the muscles, and these are called *motor processes*. (Note that talking or writing are motor output processes but that we are usually interested in the content *not* the mode of execution.)

Let's start now by looking in more detail at these functions or processes, beginning with inputs to man.

3.1   Inputs to the controller

From anatomical and physiological evidence we know that the brain itself cannot act directly—it has no muscles or glands, it can only issue instructions which are then *effected* by the bodily systems. The brain receives information via nerve impulses from an enormous variety of sources and it is constantly working, using this information to control the body, to keep it active, and to effect the things the brain 'wants' to do. Perhaps one of the most surprising things about us is how little we seem to be aware of this information flowing into the brain. Traditionally, we think of ourselves as having five senses.

What are the five senses? List them with the organ which senses them and then compare your list with the one below.

Sense	*Sense organ*	*Sensations*
visual	eye	seeing
auditory	ear	hearing
tactile	touch receptors	touching
gustatory	tongue and mouth	tasting
olfactory	'nose', i.e. receptors in nasal passages	smelling

Beyond these five, it is common to talk of a *sixth sense* when we want to refer to the intuitive or mysterious feelings or insights we sometimes experience. In fact we have many more sensory systems than the traditional five, some of which are just as important to us: our sense of balance, for example, which we use to control our position so that we can maintain any particular posture in relation to the force of gravity. Many of these other sensory systems are of vital importance in enabling us to control our actions and they usually play a significant part in the performance of skilled behaviour.

It is convenient to divide sensory information processes into those which send information to the brain about the *outside world*, and those which deal with the *internal situation* inside the body. Lists of some of the most important are given below:

Outside	*Inside*
visual	gustatory
auditory	pain
tactile	blood pressure
olfactory	head position and rate
heat	of movement
cold	joint positions
dangerous objects or	muscle tensions
effects caused by pain	body temperature

The skin is particularly rich in sensory receptors and it is possible to detect quite a number of them by simple means. One standard experiment is to chart out the different skin receptors by marking the skin with a small grid of squares, and then touching each square with a thin wire stimulator. It is possible to distinguish touch, pain and pressure sensors in this way, and if the wire probe is heated to 40°C or cooled to 1°C spots sensitive to heat or cold can be found. The hairs on the skin amplify touch sensations by means of touch receptors attached to the base of the hairs so that the hairs act as tiny levers when they are moved.

> The distribution of these sensors varies from place to place on the skin surface and you can see how close the touch receptors are by finding the *two-point threshold*. This is the smallest distance apart of two touch stimulators which can be distinguished as two separate sensations. Try with a pair of dividers or two sharp pencil points. Find a willing subject and ask them to tell you whether they can feel one, or two, points touching them. They will have to shut their eyes of course. Vary the stimulus from one point to two points randomly and start with the two points very close together. Then set the two points gradually further and further apart until the person reports reliably that there are two points. Do this at different parts of the skin. Try the back, the neck, the forearm and so on. Then reverse the procedure by starting with the two-point stimulator far apart and gradually making it closer. Draw a map of the body with the different values of the two-point threshold filled in.

The various kinds of skin receptors are shown in Fig. 3 which gives a cross-section through the outer and inner skin layers. You do not need to remember all the names of the sense receptors but it is useful to know what the receptors are sensitive to.

3.2   The sensory system

You can see from Fig. 4 that our brains have an exceedingly complex communication system which is constantly conveying information to various parts of the brain about our external and internal environment from moment to moment. A very large part of this information is dealt with at an unconscious level—most of our major sensory channels are active all the time but we are not aware of all the noises (the ticking clock?)

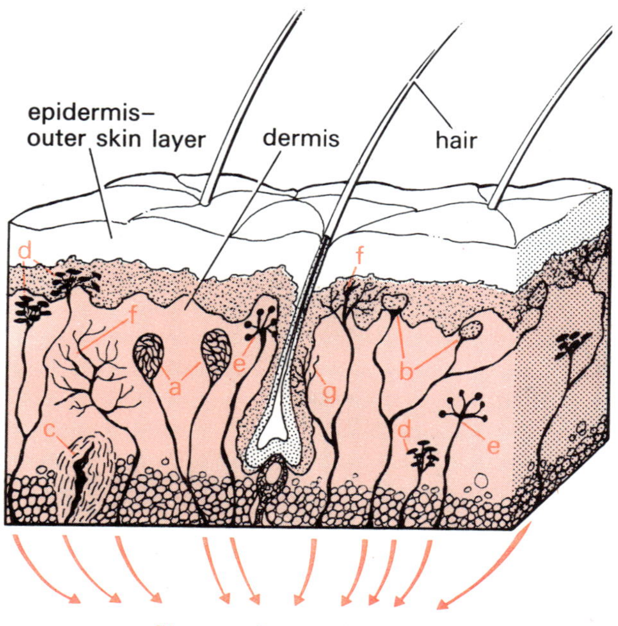

Fig. 3   Much magnified schematic view of a cross-section through the outer and inner skin layers. The various receptors embedded in the skin are tabulated below:

	receptor	sense
a	Meissner's corpuscles	touch
b	Merkel's disc	touch
c	Pacinian corpuscle	pressure
d	Ruffini ending	warmth
e	Krause end-bulb	cold
f	nerve nets	pain
g	nerve ending round hair sheath	touch (via hair)

or all the touches (from our clothes or chairs) and so on, unless we specifically turn our *attention* to them. If something unusual occurs on one sensory input, a sudden noise for example, it acts as an 'interrupt' signal and we 'notice' it. If we can account for it, say by deciding it was a car starting up in the street outside, we can 'ignore' it and go back to whatever we were thinking of or dealing with before.

If we are to link ourselves successfully with objects in the outside world it is an advantage to know something of the properties of these sensory systems. Not surprisingly, our senses evolved to suit the world as it was before man civilized it, that is our senses are suited to detect changes in the environment at the kind of levels which occur naturally. Our man-made world has developed a vastly extended range of environmental effects and now we find ourselves quite unable to cope with the levels of noise, vibrations, heat, humidity and so on which are produced by modern technology and which our modern society exposes us to or makes us work in. In many cases, as for example with noise, we can cause serious and irreparable damage to our sense organs by exposing them to excessive stimulation over a long period. The very adaptability of the human being, so useful in some circumstances, can be a source of difficulty here since we adapt quite quickly to new levels of stimulation and then do not notice them so much. This is part of the consequence of a system geared to notice *changes* and not absolute levels. Our civilized environment has also tended to limit the use we could make of our sensory systems and we just do not utilize the very wide range of detectors and sensors we have available.

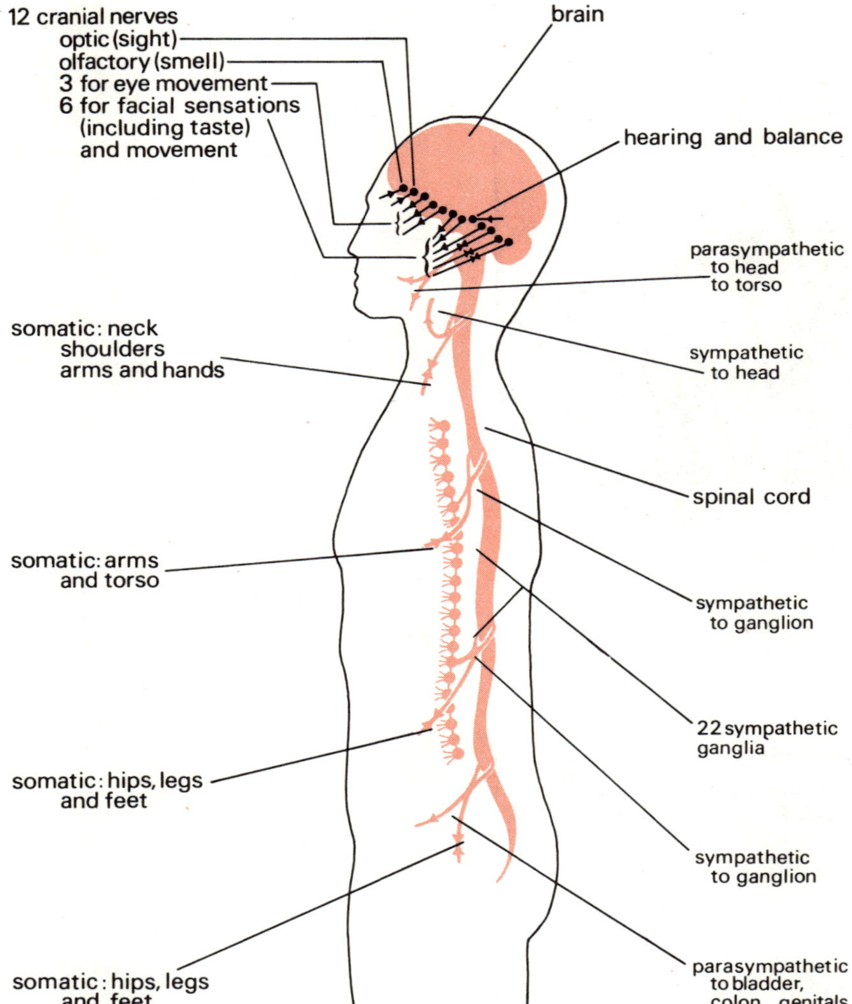

Fig. 4   Sensory and motor nerve system.

We shall now take a closer look at the more important input channels along which information flows to our brains and see how they are used in the performance of simple and complex activities.

3.3   The visual sense: the eye

This section is concerned with some properties of the human visual system and with the implications of these properties for the design of the man-made environment.

Light reaches our eyes directly from sources of light energy, such as the sun or electric light bulbs, or is reflected from or transmitted by objects in the environment. In many respects the eye is like a camera; it has a lens at the front, which focuses an image on a light-sensitive area, the *retina*, at the back. As in a camera, the image formed on the retina is a reduced, reversed and inverted one. Fig. 5 shows a schematic comparison of the optics of the eye with those of a photographic camera. The fact that in spite of the reversals and inversions we still see the world the 'right way up' is one demonstration of the ability of our brain to interpret the image on the retina and relate it to the outside world.

retina

The retina is made up of nerve cells, about 150 million of them, arranged in several layers. In the layer of *receptor cells* light is chemically converted into electrical activity which is further processed in the other layers in a way we do not fully understand as yet. The nerve impulses reach our brain

receptor cells

14

via the *optic nerve* and the *optic tract*. Fig. 6 shows the structural organization of the visual system. One important point to note is that light patterns, like all other forms of physical energy that reach our sense organs, are interpreted at the brain only after a series of complex transformations or *coding* operations at the sensory receptor and elsewhere in the nervous system.

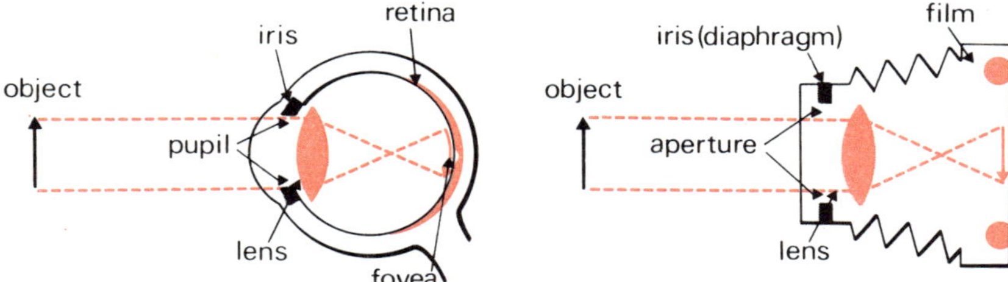

Fig. 5   Comparison of the optics of the eye and of a simple camera.

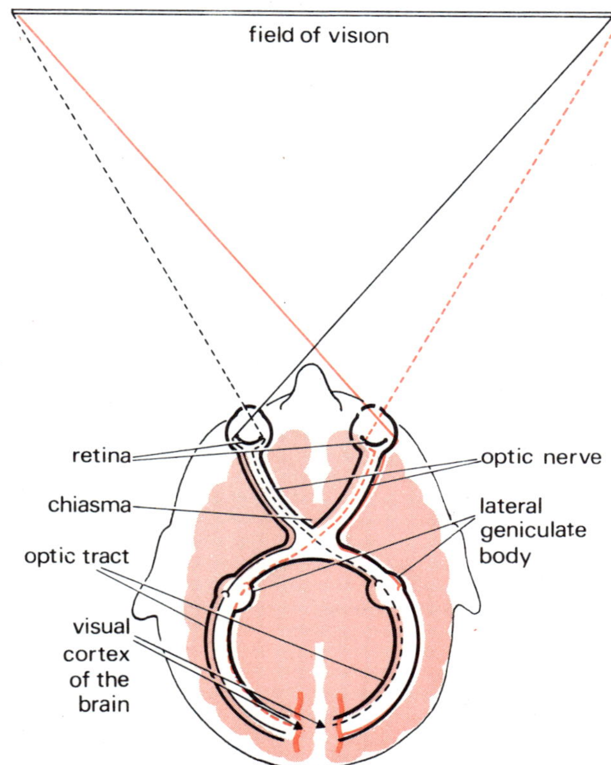

Fig. 6   Schematic representation of the component parts of the visual pathway to the brain.

Each eye can see only a limited amount of the world when looking in the directly forward position. The extent of a typical visual field, and the overlap of the visual fields of the two eyes, can be seen in Fig. 7.

Obviously our ability to move our eyes and head greatly extends our visual fields and we can cover very nearly a complete circle around us.

This simple picture of the optics of the eye is not the full story, of course. The eye has many of the additional features which are found in modern cameras, but even the most advanced 'automatic' camera is unsophisticated compared to the eye and its control systems. We have listed below some of the most important features of the eye system.

(1)   There is automatic control of (i) the amount of light reaching the retina by control of the pupil size, (ii) the focal length of the lens, and (iii) the rapid and precise movement of the eyeball itself in its socket.

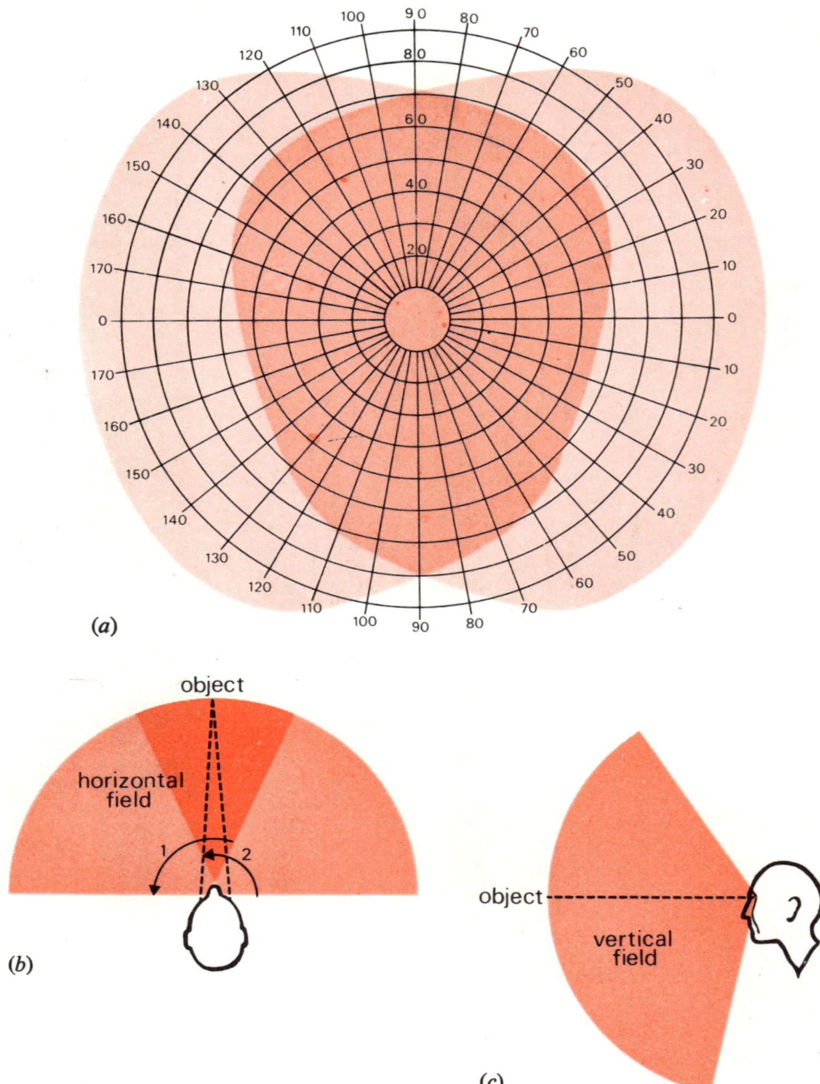

Fig. 7 (a) Field of vision of the two eyes when looking directly forward. The central portion indicates that part of the visual field which is seen by both eyes. The numbers are in degrees of angle. (b) and (c) show the field of vision in a different form.

(2) There is the remarkable ability to vary automatically the sensitivity of the eye to deal with a large range of environmental light intensities. The retina is at its most sensitive after it has been in darkness for at least half an hour. It is then able to convey a sensation of light in response to about 10 light quanta (or 4×10^{-18} joules of energy at 507 nm wavelength). This is called the *threshold of vision*. In contrast, a comfortable level of light is 100 million times as strong as this, and the tolerable upper limit is about a billion (a million million) times the threshold value. This change in sensitivity, or *adaptation*, is not instantaneous, which can be an important factor in many practical situations. For example, it occurs in driving a car into and out of an underpass in daytime, or in trying to spot hazards at night while the headlights of oncoming cars keep changing the light levels and hence the sensitivity and detecting ability of the eye. You will have noticed the same effect when you go into a darkened cinema from daylight. At first you cannot see anything other than the screen but slowly the rest of the cinema becomes visible.

(3) There is a high degree of coordination called *convergence* between the two eyes to keep an object of interest properly imaged on both retinas, whatever its distance or direction is. As we shall see later, you can see

threshold of vision

adaptation

convergence

16

visual detail best when the image falls on a small part of the retina near the centre called the *fovea* (see Fig. 5).

fovea

Try testing for convergence by observing how a person's eyes follow a pencil as it is brought up close to the nose.

(4) Finally in this list of properties of the eye, there is the sensitivity to a range of frequencies in the electro-magnetic wave spectrum, extending from about 400 nm (violet) to 800 nm (deep red). Fig. 8 shows the 'band-spread' of visible light compared with the full range of the electro-magnetic radiation spectrum. Notice the very narrow range of wavelengths to which the human eye responds. The effect of light on a human observer depends on its frequency and amplitude.

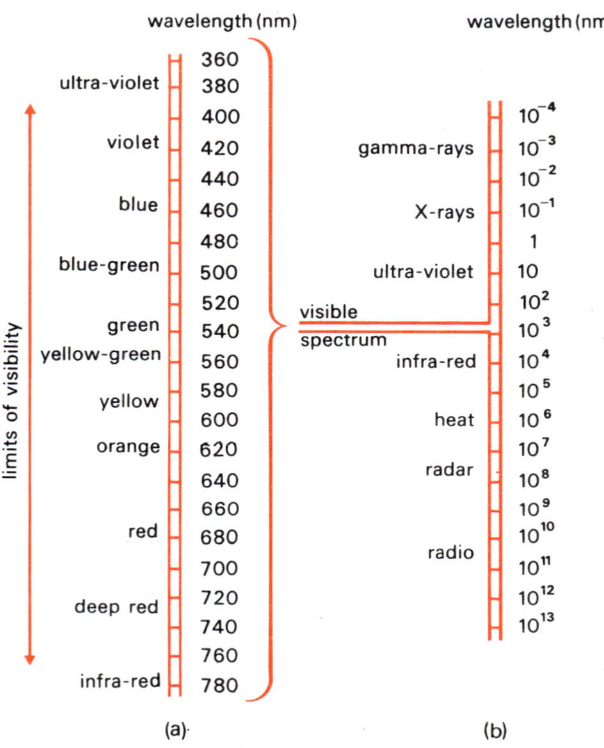

Fig. 8 *(a) The visible spectrum expanded. (b) The electromagnetic spectrum.*

The sensation is dependent on intensity level: at low light levels colour tends to disappear or change shade; flowers, for example, appear to change their colour towards dusk.

The colour observed depends on the frequency or combination of frequencies observed and the brightness depends on the amplitude of the vibrations. The effect of white light is produced when the complete spectrum (or at least the greater portion of it) is observed simultaneously. The human eye is, in fact, more sensitive to some wavelengths than others (see Fig. 9), with maximum sensitivity in the green–yellow region.

Fig. 10 shows the spectrum of some common light sources which produce effectively 'white' light compared with the discrete emissions produced by a mercury vapour lamp.

All these remarkable properties of the eye come into play, *usually at the subconscious level*, for most tasks involving vision. They all contribute to the visual skills we have listed earlier, so let us look at these skills in greater detail.

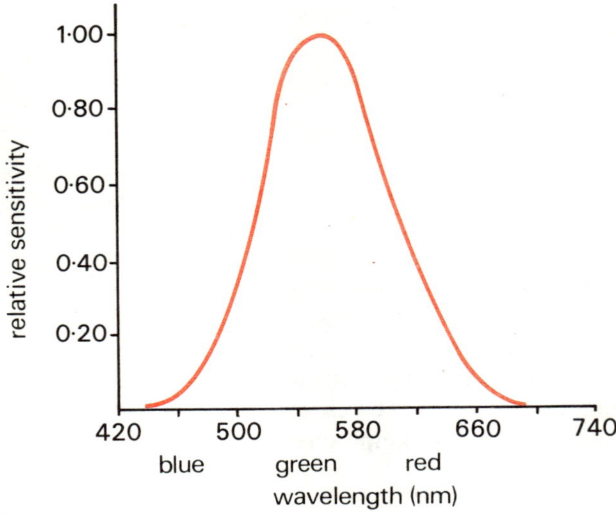

Fig. 9   *The sensitivity curve of the eye.*

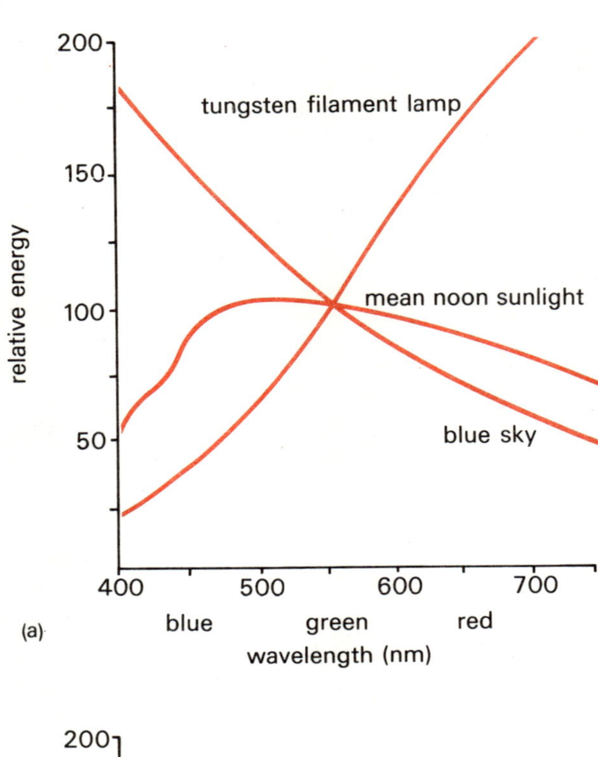

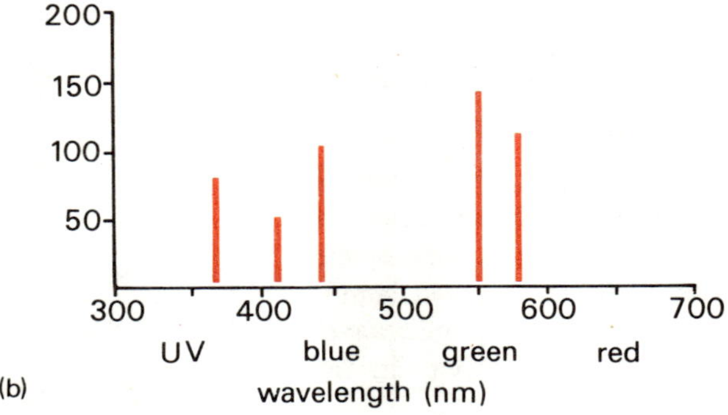

Fig. 10   (*a*) *Spectral energy distribution curves from sunlight, light from a blue sky and light from a tungsten lamp, all of which will be seen as effectively 'white' because of chromatic adaptation of the eye.* (*b*) *Spectral distribution of light from a low-pressure mercury vapour discharge lamp.*

3.3.1 Brightness discrimination

Everyday experience shows that different parts of the environment create different impressions of brightness. Why does this happen? Clearly an important factor is whether the light is coming directly from a source or if it is reflected. If it is reflected, differences in brightness could be due either to differences in the brightness of the source or in the reflective properties of the surface of the object, or both. In both cases, light may be lost (by absorption) in the medium between the observer and the source or object. We use these brightness differences to gain information about the outside world.

> Pause here and write down three situations in a car driver's experience where differences in brightness provide significant information for him.

In doing this, it has probably become apparent to you that the ability of the eye to discriminate fine detail in the visual environment is of great importance. We rely very much on our eyes to serve as an 'early-warning system', and it is obviously important to get accurate information about objects at a distance, but the image formed on the retina gets smaller as the distance between the eye and the object being viewed increases. Intuition and experience tell us that there must be some size limit, related to the properties of the eye, below which fine detail is lost. This size, however, is not the *actual* size of the object (its physical dimensions) but its apparent size, that is the size as projected *on to the retina*. This retinal size is usually measured in terms of the visual angle subtended by the object at the eye. The idea of visual angles is illustrated in Fig. 11. As the distance n is about 15 mm, a 2·5 mm long object such as a capital letter on this page, at the usual reading distance of 25 cm will form a retinal image only 0·15 mm long or a fraction of the size of a full stop. You can see that an object near the limits of size for detection could be a very small object quite near to the eye, or a large object a long way away.

The capacity to discriminate fine details of objects in the visual field is called *acuity*. It is defined as the *reciprocal of the visual angle of the smallest resolvable detail of an object when this angle is expressed in minutes* (1 minute is 1/60 of a degree). Fig. 12 illustrates the concept of acuity. It is the familiar eye test chart reproduced here one-quarter of its actual size. (Fig. 12 is on p. 45 of this unit. Remove the page if you wish.)

With this chart you can perform an informal test of your own visual acuity. Prop the chart up vertically about 6 metres away from you. At this distance a person with normal vision should be able to identify the three letters in line 3 of the card. Acuity is affected by, among other things, intensity of illumination, so your test should be carried out at normal reading light intensities.

> Calculate your, or another person's, visual acuity by choosing a line of letters on the chart in Fig. 12 and moving towards or away from the chart to a distance (D) where all the details of the letters in the line are just visible. Measure the line width of the letters of interest (w), then calculate the visual acuity from the formula in Fig. 11. If you have time, repeat with letters of different size.

Clearly the performance of most visual tasks depends on acuity. These include:

detection (whether or not an object is present in the visual field)

recognition (the identification of seen objects or their parts)

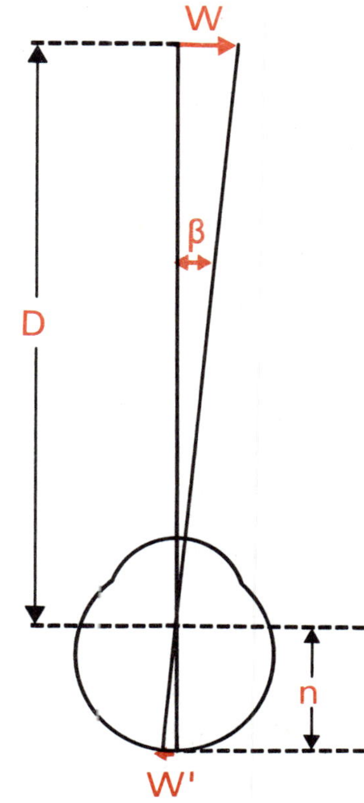

Fig. 11 *Diagrams to illustrate the specification of visual acuity. The width, w, of the test object is so small that it can just barely be discriminated at a distance, D, from the eye. The width (or other appropriate critical dimension) of the object so discriminated then subtends an angle β at the eye. This angle, provided it is small, is given approximately by the relation $\beta = w/D$ in radians, or $\beta = 3450w/D$ in minutes of arc. Visual acuity is defined as the reciprocal of this angle in minutes (i.e. $D/3450w$). The size of the object on the retina is given by w' ($w' = wn/D$).*

resolution (response to separation between parts of seen objects)

localization (the detection of small displacements of one part of an object with respect to its other parts).

In a detection situation a single dark line against a bright background, such as a fine crack in a white china plate, can be detected under favourable conditions when the width of the line subtends about half a second of visual angle at the retina (1 second is 1/60 of a minute). The amount of light on the background is of great importance in this task: if it is too low, the reduced contrast makes detection more difficult, while high levels of background lighting give rise to glare and a consequent decrease in detection ability.

The other three tasks listed above are similarly affected by the differences in the absolute and relative amounts of light on the object and its background.

In the resolution task, a series of parallel alternating black and white lines of equal width can normally be *resolved*, that is each line can be seen separately, when the width of the lines subtends an angle of at least half a minute. Compare this with the minimal width of a single dark line as used in the detection task.

In the localization task, acuity is often measured by the ability of a person to detect misalignment of two straight lines which, when aligned, form a continuous straight line. Those who have tried to user vernier scales will understand why this ability is referred to as *vernier acuity*. People with normal sight can report misalignment when the two lines are displaced by at least 2 seconds of visual angle. This corresponds to about 0·01 mm when the lines are at a distance of 1 m from the eye.

All the above data on acuity refer to situations when the image of the object falls on the most sensitive part of the retina, the fovea. When the image falls outside this area (for example, when you 'look' out of the corner of your eye) by 1 degree or more of visual angle, the resolving power of the eye drops to some two-thirds or less of foveal vision. On the other hand, your ability to *detect movement* and flickering effects in the visual field is much better in these non-foveal parts of the retina. You can see this if you look slightly away from a TV screen or fluorescent light. When the image falls on the outer regions of the retina you can often see flickering, but if you look straight at the source, the flicker disappears.

These data on brightness discrimination can be summarized as follows:

The stimulus must be large enough in extent, it must be distinguishable from the background, and it must provide sufficient light to be detectable.

The duration of a stimulus is also an important factor in the visual effect it causes; it must be present long enough for the eye to react. If the illumination is intermittent, the effect is dependent on the frequency at which the illumination is increasing and decreasing. When the frequency is below about 10 cycles a second, each change is seen separately. Between 10 and 20 repetitions a second, depending upon the individual, a flicker effect is observed, which may be quite uncomfortable and disturbing.

This phenomenon can sometimes be experienced when driving a car. If you happen to be driving along an avenue of regularly spaced trees or fence palings with the sun shining at you between them, at a particular speed the flicker effect is produced and it may become strong enough to distract your attention from your driving.

When the frequency of light variations exceeds about 20 a second, flicker gives place, in most of us, to an impression of uniform illumination. This impression is so strong that we are not usually aware that, for example, a fluorescent light is actually switched on and off 100 times a second by the alternating voltage of the mains, or that television and cinema pictures are, in fact, presented one frame at a time at 1/20 to 1/25 second intervals.

Unusual visual effects occur when not only the illumination is intermittent, but the object itself contains a repetitively moving part. When the rates of repetition coincide or are an exact multiple of one another, the object can appear to be stationary. This is called the *stroboscopic* effect and can be both useful and dangerous. It is useful because we can observe the behaviour of a rapidly rotating or cyclically moving object quite easily. We simply arrange for the object to be lit by a flashing light and we adjust the frequency of the light flashes so that they always occur when the object is in the same relative position. It then appears to be stationary and we can study it closely to see what is going on. The danger arises when this happens accidentally and unexpectedly. A lathe or drill chuck, for example, rotating at 6 000 rpm will be turning at 100 times a second. If the scene is lit by a light which switches on and off 100 times a second, the chuck can appear to be completely stationary. In a busy workshop the noise of the machine is lost in the general row and it is all too easy to reach out to touch the apparently stationary chuck with serious results. For this reason workshops are often lit by special banks of lights *out of phase* with each other so that at least one light is always on at any one time.

Note that when the repetition rate of movement of the object and of the illumination is only slightly different, an apparent slow movement effect is seen. A good example is the peculiar way stage-coach wheels speed up and slow down when they are seen on film or television.

3.3.2 Colour discrimination

As with the detection of changes in brightness, the eye is much better at detecting *differences* in hue than at absolute determination of colour. Indeed, in the normal eye, hue discrimination is such a sensitive process that it is possible to notice a difference in hue without being able to state what the different colours are. It is important to remember, however, that variations in the amount of light can create the illusion of changes in hue. However, blue, red and purple tend to retain their hue more than do other colours as the light intensity falls.

The discriminating ability of the eye for hues is best at around 480 nm (blue) and 580 nm (yellow).

The colours red, yellow and blue are usually referred to as the three principal or primary colours.

A fascinating property of colour vision is that the observed colour of an object is dependent on the colour of light which illuminates it. For example, a blue object illuminated by yellow light appears black to the normal observer. There are, however, exceptions to this rule. If the blue object happens to be a car, for example, and the yellow light is the sodium light used for street lighting, then the car will still appear to be blue to its owner, or to anyone familiar with it. This phenomenon is called colour constancy, and we know little about its cause as yet.

About 8 % of men, and a much smaller fraction of women, have abnormal colour vision in that they require a different mix of primary colours to match a particular colour than do normal individuals. These people tend to confuse red and green or red and blue or green and blue colours. A very few people suffer from what appears to be a complete lack of colour sensitivity.

3.3.3 Space perception

In space perception we come across one of the most interesting aspects of vision. The retina is a two-dimensional surface, yet we observe the world as having not only height and width but also *depth*. We are also able to estimate distances fairly well. How we see in three dimensions is another aspect of vision that is not fully understood, but it is known that the three-dimensional view of the world around us is formed in the brain. You may be able to appreciate the problem better if you think of the two-dimensional surface of the television screen as the two-dimensional window through which a three-dimensional world is observed.

> Although you may not be conscious of all the properties of a television picture that leads you to accept that the two-dimensional picture is a true representation of the three-dimensional world, try to compile a list of the factors which you think help us to form impressions of depth. Compare your list with ours, which is set out below.

We have drawn up a list in two sections: one with the cues for which only one eye is needed and the other for those where both eyes are used.

Cues supplied by one eye:

(1) the relative size of the objects at the retina (small retinal image—distant object);

(2) perspective (parallel lines appear to converge at a distance);

(3) aerial perspective (when the visual detail on a surface is small, it appears to be far away);

(4) interposition (with overlapping objects the overlapped one appears to be further away);

(5) parallax (if the eyes are moved from side to side, near objects appear to move *against* the direction of actual movement in relation to more distant objects).

Cues which can only be provided by both eyes include:

(1) convergence (when a near object is observed with both eyes, the eyes turn slightly inwards);

(2) stereoscopic images (the two eyes receive and combine slightly different images of the same object in space).

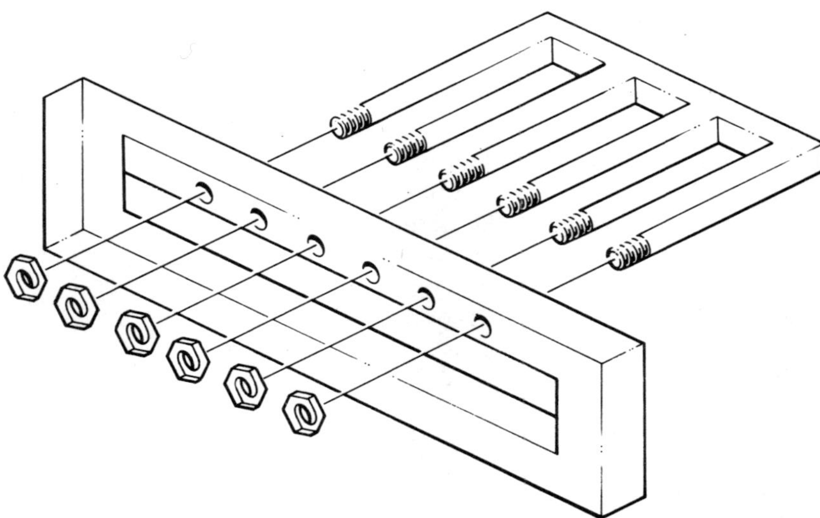

Fig. 13   A drawing of machine parts? Impossible! The illusions are created by flaunting the conventions of representing three-dimensional objects by two-dimensional drawings.

Normally all these visual cues are used together to provide an observer with a good estimate of relative position. Confusion, however, can be generated if we artificially give conflicting cues; a famous 'impossible set of objects' is shown in Fig. 13.

3.3.4 Form perception

Visual skills that we have looked at so far provide us with information on the existence, location and colour of objects and other parts of the environment, but a major task has not yet been touched upon: the actual identification, recognition or naming of objects.

One of the most important and fascinating aspects of vision is our ability to recognize *patterns*. The brain and the eye together can recognize any of the following as an A:

No machine is yet capable of doing this and how we do it is the subject of much research at the moment. One interesting aspect of this is that we acquire these abilities at a very early stage in our development; young children have to learn to 'see' the shapes and patterns in the world around us and we do it so successfully that we tend to accept this ability as a fundamental property of the visual system instead of a learned one. One important property of any form is its contour or outline. There can be no doubt that contours play a significant role in form perception.

The process of form perception is prone to *illusions*, where the subjective impression gained of a situation is different from the actual, objective situation. A familiar illusion is illustrated in Fig. 14.

illusions

Fig. 14 *The vertical–horizontal illusion.*

On the left the vertical line appears to be longer than the horizontal, although the two are of equal length, and on the right, the horizontal side of the square appears to be longer.

Many illusions are caused by incomplete information, such as in representations of three-dimensional objects in two-dimensional pictures, or by conflicting information provided by the object and its background. Conflict between visual perception and body position and motion sensing is also a frequent source of illusion. You may well have been fooled into putting on the brakes of your car while stationary in a traffic queue when cars going the same way in the next lane started to move past, leading you to believe that your own car was rolling backwards. Pilots often report conflict between the reading of their position and direction-sensing instruments and their own sensations as to what the plane is doing. It is interesting that in most such cases we are far more likely to believe our eyes than our other senses, even if, in experimental conditions, the eyes are given the 'wrong' information.

3.3.5 Movement perception

Movement is perceived by the visual sense in two situations:

(1) when the object or the observer is physically moving (real movement), and

(2) when successive views of a scene are presented for a duration anywhere between a few milliseconds to about 400 milliseconds, with the object in a new location in each view (apparent movement).

Our ability to detect movement is strongly influenced by the presence of stationary (reference) objects in the field of view. When such objects are present, movement at a rate of 1–2 minutes of visual angle per second can be detected, but without them movement needs to be at 10–20 minutes per second to be seen.

Notice again that relative judgements can be made more easily than absolute ones.

3.4 The auditory sense

We have spent rather a lot of our space on the visual sense because in many ways it is the most vital sense as far as everyday living and working is concerned. Blind people are much more seriously handicapped in industrial situations and in man–machine interactions than are deaf people, for example. Deaf people are, however, to a great extent deprived of everyday opportunities to communicate with other people and this can be a serious loss indeed. If deafness occurs before or during childhood, the results can be very damaging for the development of personality and the deaf person can suffer more than the blind person from this isolation from social contact and the use of language. It is worth noting that many animals have poor vision but manage to survive very well because of sensitive auditory and olfactory senses.

Two reasons why hearing is important are:

(1) Initial communication and hence understanding arise primarily from learning spoken language and building up symbolic thinking processes.

(2) Hearing is essentially a non-directive sense. We cannot see something unless we look at it, but we *can* hear things from anywhere around us. We are even 'listening' when we are asleep.

Example 1

Why do you think this second property of hearing is so important?

For answer, see end of unit.

3.4.1 Sound

Our auditory sense responds to sound waves rather as the visual sense responds to light waves. Unlike light, however, sound needs a medium to travel through or along. In fact sound is just the name we give to vibrations of particles of a substance over a certain range of *rate of vibration*. The rate of vibration is called *frequency* and we can 'hear' vibrations in the range from about 20 vibrations a second up to a top limit of 20 000 vibrations a second. These vibrations per second are called hertz, so the normal range of human hearing is about 20–20000 Hz but it is normally only the very young who can hear at the upper limit. For most of us, sounds above 12–15 kHz* are inaudible.

frequency

* 1 kHz = 1 kilohertz = 1 000 Hz.

Virtually all of the sounds we hear come to our ears, the receiving sensors, via air pressure waves, so we shall concentrate on these, but it is worth pointing out that we can get sound sensations directly to the sensors inside the ear via bone conduction. Since sound waves travel at different velocities in different materials this can be an advantage. For example, sound travels four times as fast in steel as in air so you could 'hear' a train coming by pressing your ear to a rail long before the air sound waves reached you. Another advantage is that if you cannot receive the air sound waves you can still often hear via bone conduction; you can test this by touching a vibrating source with your teeth.

3.4.2 The ear

Sound waves travel in air at 340 m/sec (at sea level) and arrive at the outer ear (called the pinna) which acts to some extent to concentrate the sound and funnel it down the ear passage. The pressure wave comes up against the ear drum which vibrates as the wave does. These vibrations are transferred by the mechanical motion of three small bones to the end of the auditory sensing device, called the *cochlea*. The cochlea, which is shown in the diagram in Fig. 15, transforms the mechanical vibrations into nerve impulses which travel to the brain.

cochlea

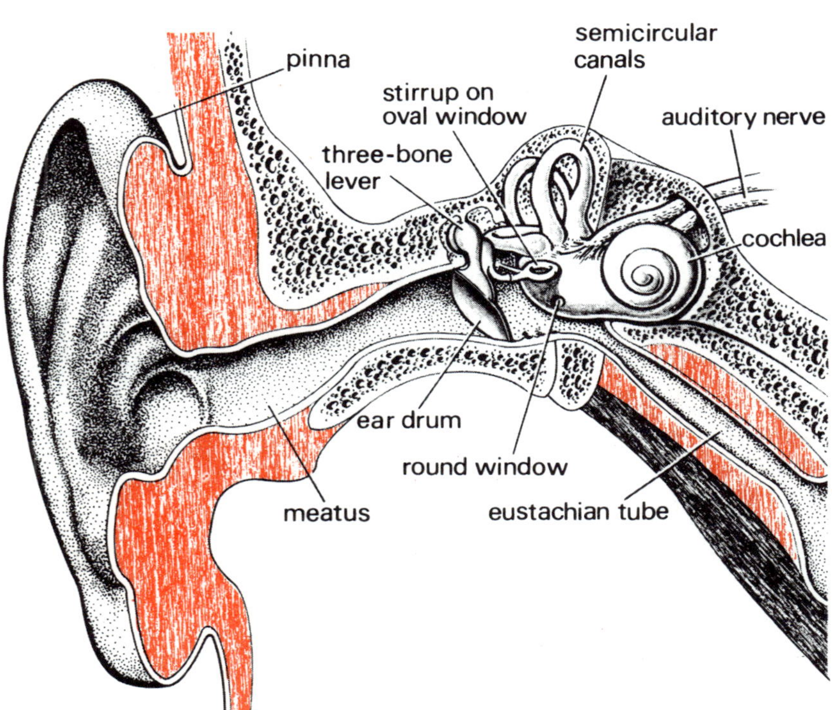

Fig. 15 The ear mechanism. The sound waves enter the cochlea via the oval window and are then transformed into nerve impulses. The round window allows the pressure wave in the cochlea fluid to dissipate. The diagram also shows the semi-circular canals which sense motion of the head, and the eustachian tube which connects the space behind the eardrum with the mouth cavity.

Sound waves are like many other kinds of wave and they are basically vibrations of particles of matter. The general properties of vibrating masses or particles are of fundamental importance in engineering and you will be meeting this topic again in later technology courses. At this stage the two important things to know about waves are: they have a frequency of vibration as we have already said, measured in hertz, as so many vibrations a second, and they have an *amplitude* which is a measure of how far (or how much) the particles vibrate in each cycle. Typical single frequency sound waves are illustrated in Fig. 16.

amplitude

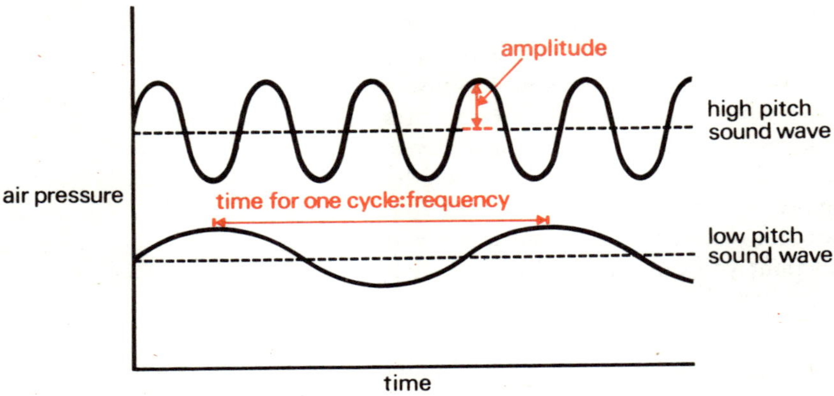

Fig. 16   *Typical single frequency sound waves.*

We very rarely come across single frequency waves in practice and most sound waves are complex mixtures of different frequency waves like those shown in Fig. 17. However, in laboratory experiments it is usually easier to use single frequency waves.

The ear responds to both different frequency and different *amplitude* sound waves. The amplitude of a sound wave is a variation in *sound pressure* and so we measure it in pressure units, e.g. Newtons per square metre*. We have already mentioned the frequency range (although this depends very much on age as we shall see later) but what about pressure? There is a lower limit below which we cannot hear sounds and this is an advantage, otherwise we might be able to hear random movements of the air particles around us and this could mask other more useful sounds. This lower limit is called the *absolute threshold of hearing* and varies a little from person to person. An average figure can be found by making some very faint sounds of different pressure and seeing how many of a large group of people can detect them. Because the threshold might be expected to change with frequency, the experiment is repeated at a number of different frequencies across the auditory range. These tests are carried out on individuals in special soundproofed chambers. The graph in Fig. 18 shows the results of some experiments on typical Americans using single frequency waves. The frequency of the sounds is plotted on the horizontal axis and their pressure on the vertical axis. The number of people in the test group who could hear each sound at the different levels is shown as a percentage, so that the bottom line shows the levels which only 1 in 100 people could detect, and the top line the levels which 99 in 100 could reliably report. The figures usually quoted are those which 50% of the population can reliably detect. Notice that the absolute threshold changes with frequency and that, in general, our ears are most sensitive at about 4 000 Hz.

When the sound pressure approaches 20 N/m² we feel a tickling sensation and discomfort in our ears. This upper limit is called the *threshold of feeling*.

3.4.3   Objective and subjective qualities

The characteristics of a sound wave—the frequency or frequencies present, and their amplitudes and relative phases—are physical qualities which can be measured with appropriate instruments. The ear and the brain, however, do not hear these physical qualities as such, but as subjective experiences just as light of wavelength 505 nm is seen as the colour *green*. Subjective experience is, of course, unique to each individual but we can study it and it turns out not surprisingly to be similar for most of us.

---

* The Royal Society has now adopted the pascal as the SI derived unit for pressure, replacing newtons per square metre.

Different frequency sounds are heard as having a different *pitch*, and different pressure levels as having different *loudness*. So *pitch* is our subjective assessment of frequency and *loudness* is our subjective assessment of pressure. The subjective experiences are not exactly the same as the physical quantities and it is worth looking at the differences in a little more detail.

3.4.4 Frequency and pitch

Single frequency sounds from 20 up to 20 000 Hz are each heard as differently pitched sounds and we can usually tell the difference between a sound of 200 Hz and 205 Hz (this is *not* the same as saying that we can *identify* all the different frequency sounds in the audible range). The pitch we hear, and the actual frequency of the sound, are fairly well matched but the pressure of the sound makes a difference. For sounds below 500 Hz the pitch sounds lower as the pressure increases. So sounds at these frequencies and of *high* pressure have a lower pitch than when their pressure is *low*. In a similar way, high sounds above 4 000 Hz seem higher in pitch if they are of high pressure. The *difference* between two sounds of different frequency is also heard differently depending on where the difference occurs in the frequency range. The two sounds 256 Hz (middle C)

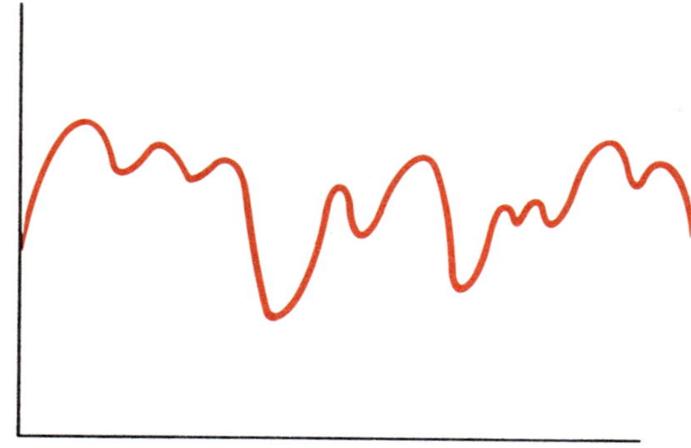

Fig. 17 *Complex wave.*

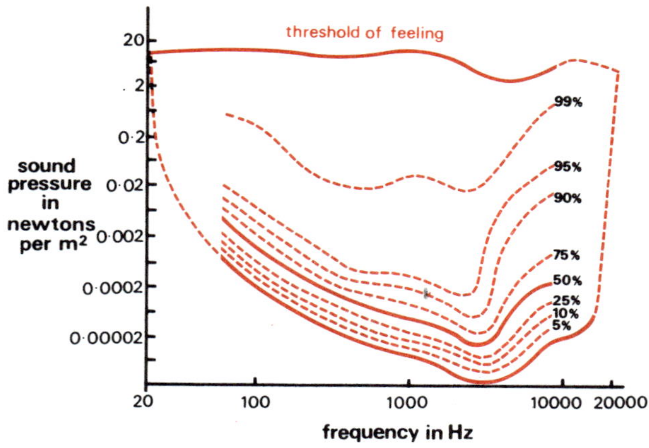

Fig. 18 *Sensitivity of the ear to sounds of different frequency. Curves are shown for the percentage of the test sample who could just hear each different frequency tone.*

and 512 Hz (the next C above middle C) are an octave apart, the higher frequency being twice the lower one, but they are heard as being further apart than either 32–64 Hz or 2048–4096 Hz although both these pairs are also one octave apart. So a doubling of frequency does not necessarily mean a doubling of pitch. This aspect could be important where a man doing a task has to judge the differences between sounds, since he would underestimate low or high pair differences compared to middle of the range differences.

Earlier on we mentioned the effect of age on your ability to hear different frequency sounds. You have probably come across the dog whistle which dogs can hear and we cannot, and if you are over, say, 40 or even 35, you may have noticed (or not noticed) that the annoying whistle heard with most TV sets has lessened or even disappeared altogether. But ask your children or younger brothers or sisters whether they notice it still and you may be surprised at their comments. The reason for this is shown in Fig. 19 and you can see how our sensitivity to high frequency sounds falls off with increase in age.

We have already mentioned the change in threshold with frequency and it is interesting to note that our ears are most sensitive at about the frequency of a piercing scream. It may be that this is no accident and it could have been an advantage for primitive man to be able to detect readily sounds like this which signalled danger. Since, of course, sound waves spread out through the air and get weaker the further they travel, this would mean that sounds at these frequencies could be heard from further away.

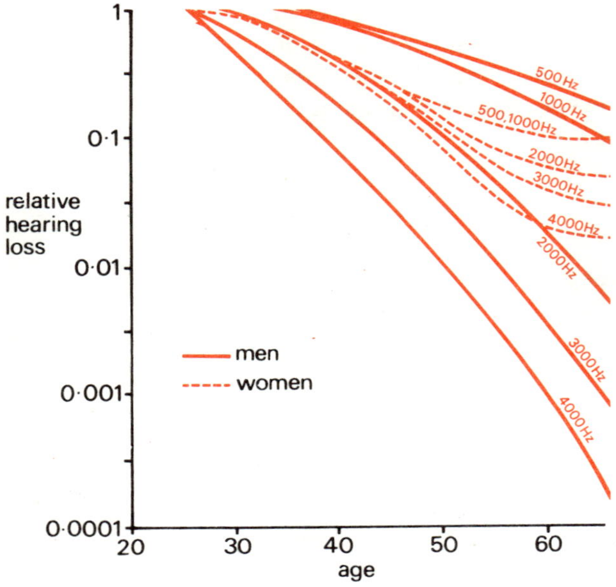

Fig. 19 Curves for hearing loss with ageing for men and women. The graph shows the hearing loss expressed in relation to the sensitivity at 25 years to a 4 000 Hz tone. For example, the hearing sensitivity of a 50-year-old man for a tone of 4 000 Hz is about 1/100 of his sensitivity at 25 years.

3.4.5 Sound pressure and loudness

The range of amplitudes we can hear is very large, from pressures of 0·00002 up to 20 N/m² , and the relation between pressure, the physical quantity, and loudness (the subjective effect) is interesting. If we start with a low pressure sound and then increase the pressure until it sounds 'twice as loud' we find the actual pressure has been increased approximately threefold.

How loud a sound seems also depends on its frequency, and a sound wave of 1 000 Hz at 0·002 N/m² pressure sounds 3 times as loud as a 100 Hz wave at the same pressure. Experiments have been carried out to find out how loudness varies with frequency and the results of a typical series of tests are shown in Fig. 20. Each individual curve shows a line of 'equal loudness' revealing the pressure a sound has to have to be heard as the same loudness. Note that the lower curves follow a similar pattern to those for the threshold of hearing curves (Fig. 18) and it is probably only to be expected that all the very faintest sounds we hear will sound equally loud, or rather equally soft.

You can also see that the loudness of lower frequency sounds is rather less than that of middle range ones at the same pressure, but notice how this effect increases as the overall sound level falls. This means that at low sound pressures, low frequency sounds appear much less loud than sounds in the range 1 000–5 000 Hz at the same pressure. The practical effect of this is that when you are listening to a sound source which is emitting a wide range of frequencies (such as a good 'hi-fi' set), you will tend to lose the low frequency sounds as the overall pressure falls. Some hi-fi amplifiers now have a 'contour' control which automatically produces an extra amplification of low frequencies at low volumes to offset this effect. This is why music on the radio tends to sound shriller as you turn the volume down and why hi-fi enthusiasts turn the bass control up as they are forced by their neighbours to turn the volume control down.

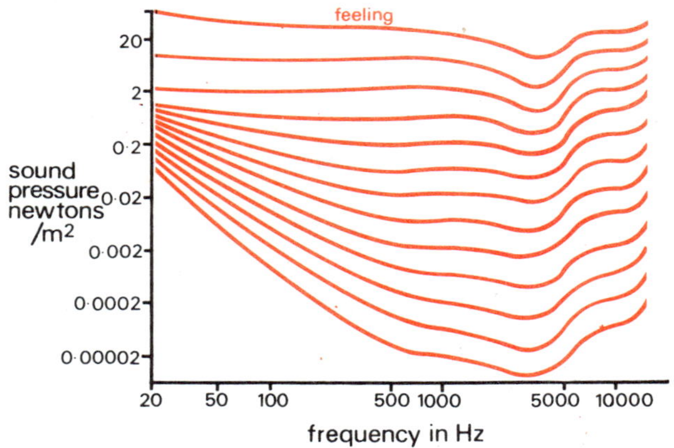

Fig. 20   *Equal loudness level curves. Each curve represents an equal loudness level.*

3.4.6   Locating the source of sound

So far we have talked about the properties of an *ear* but not of the *ears*. Because we have two ears we can do things we could not do with one. This two-ear system is called the *binaural* system and it is similar to binocular vision. Binaural hearing enables us to compare in the brain the sounds which arrive at each ear and one of the main uses of this is in the location of sound sources. Human beings and animals are surprisingly good at being able to tell where a sound has come from.

binaural system

> Try a simple experiment: get a friend to sit in a chair and close his or her eyes. Ask him to sit facing in one direction and not to move his head at all. Use two coins, or something similar, to produce a sharp click sound and make clicks at various places around him about six feet away. Obviously, you will try not to give away any clues about where you are and it is worth getting several people to stand around the room to help you if you can. Get the subject to point to the apparent source of the click and see how good he is at locating the sound. Make a particular note of his performance in the medial plane (see Fig. 21).
>
> Is the subject better when the click is repeated two or three times in succession? Use a ticking clock, for example. When repetition is used, compare the head still situation with one where you allow the person to move his head if he wants to. Does this improve location?

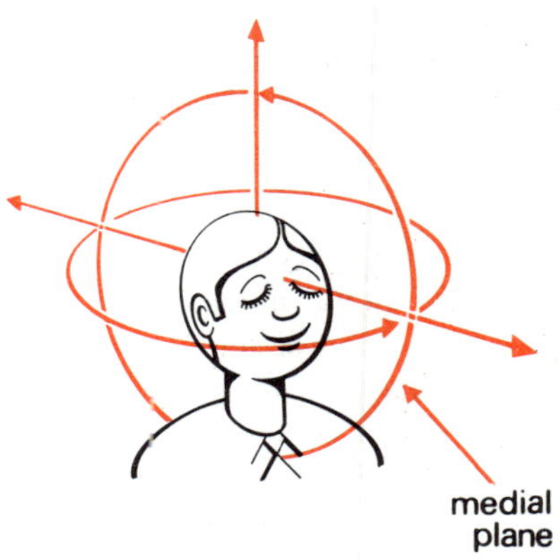

Fig. 21.

The results should reveal one important factor in location: it is accurate when the sound source is either nearer one ear or the other and less accurate at points equidistant from the two ears.

How do we use the two-ear information to locate sound sources? There are two possible mechanisms involved: if the sound source is to one side of the head the nearer ear will hear a slightly louder sound. This is accentuated by the fact that the further ear is usually in a kind of *sound shadow* caused by the head itself. This difference in loudness or intensity is really only marked with high frequency sounds above about 3 000 Hz because lower frequency sounds have longer wavelengths and they can travel round the head more easily, thus lessening the loudness difference. The brain uses this very small intensity difference to locate the direction the sound must have come from.

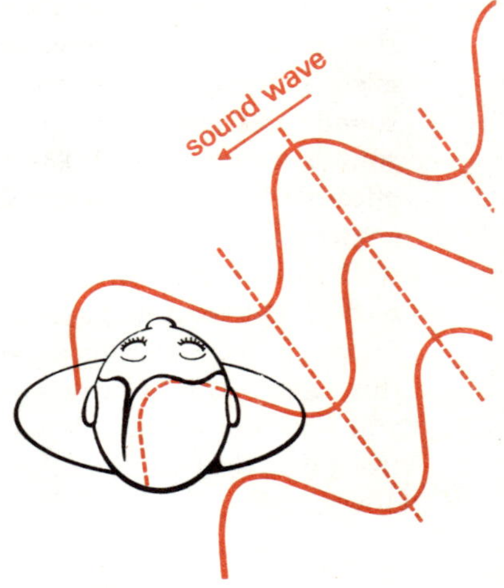

Fig. 22.

With lower frequency sounds the brain may use a different technique based on time, or more strictly, *phase* differences. A sound wave coming from one side of the head reaches the nearer ear sooner and by the time the same wave reaches the other ear, the nearer ear will be hearing a later part of the wave (Fig. 22). This effect occurs because the sound wave takes an appreciable time to travel the extra distance from the nearer ear to the further. 'Appreciable' here means more than 0·000 065 seconds or 0·065 milliseconds. (A millisecond = 1/1 000 second.) The brain is capable of detecting these very small time differences and of using the comparison between the two sounds at each ear to decide on the direction of the sound wave, so the brain is performing a remarkable time difference sound comparison analysis.

Naturally the effect is greater when the source of sound is at 90° to the head but it is still quite large at 45°, as Table 1 shows.

Table 1

| Direction angle | Time difference in milliseconds | |
	Source close to head	Source distant
0°	0	0
45°	0·40	0·38
90°	0·80	0·65

It would seem that the brain uses the time difference technique to locate the source of lower frequency sounds below about 1 000 Hz, but the effect diminishes below 200 Hz and sounds below this frequency are difficult to locate. At these very low frequencies the difference in the two samples of sound wave arriving at each ear becomes too small to be used. So two different possible sound location techniques are used, intensity differences primarily for high frequencies, greater than 3 000 Hz, and time differences for lower frequencies, in the range 200–1000 Hz.

Did you notice any improvement in location in your experiment when your subject was allowed to turn his or her head towards the sound source? The ability to locate a repeated or continuous sound is probably helped by slight movements of the head because the subject can use the *change* in binaural sound to home on to the right direction. This is probably an important factor because it is usually easier to tell if something is getting bigger or smaller (such as a loudness *difference*) rather than trying to judge absolute magnitudes of the stimulus. Many animals, of course, can actually move their outer ear to scan for the sound source. The pinna is useful here even in human beings because it causes a slightly different sound to be heard when the sound source is behind rather than in front of the head. This means that some location in the medial plane is possible.

It is not enough to know the direction of a sound source, i.e. its *bearing*, to locate it. You would normally want to know its distance as well. Probably we simply use hearing to orient our head so that we then have our eyes pointing in the direction of the sound and can see the source itself. However, we can do some distance sensing from the nature of the sound quality alone. We know something already about the change in relative loudness of different sounds at different intensity levels, so a sound source can be compared with a remembered pattern and if, for example, we find the lower frequencies rather quieter than we expect, we might conclude that the source was far away. Other aspects are undoubtedly used which we still know very little about.

You may have noticed that there is a gap in the sound location system and it is in fact hard to locate sounds around 2 000 Hz, and strangely this is in the frequency region where the ears are most sensitive. It may be no accident that some animal warning cries tend to lie in this poor location region. Presumably an animal spotting danger wants to warn others of its kind without itself being located and this is clearly a good region in which to have sensitive hearing as well.* Mating calls, on the other hand, tend to be made at frequencies where location is good; should we be surprised?

3.5   Other senses

The other main senses which were listed in Section 3.1 have been discussed briefly earlier; we have looked at some of the skin senses and you can find out more about the olfactory and gustatory senses from the references given at the end of this unit. One group of senses, however, deserves more attention here. These are the signals which come from *within* the body structure and provide us with information about the position of the body. There are two basic sensory devices within the body to tell the brain about body condition: those attached to the skeletal muscles and tendons which signal back to the brain the state of tension or relaxation in each muscle; others sited at the joints between bones indicate to the brain the angles between the bones and the rates of relative movement.

From these two sources the brain can work out reasonably well where all the parts of the body are at any time, even when it cannot 'see' the limbs. Additional information about the orientation of the head comes from the semicircular canals behind the ears so that even with your eyes shut you can tell to some extent what position you are in.

---

* Animals do, of course, have rather different auditory characteristics, so these regions might vary from species to species.

Human judgements

The preceding sections have set out some of the basic properties of the sensory system but the brain does not use this information in quite the way you might expect. For example, you can probably tell the difference between a sound of 1 000 and 1 002 Hz, if they are played to you one after the other, and also between 1 002 and 1 004 Hz and so on. This ability to detect a difference of 2 Hz is not a constant property; at higher frequencies the difference has to be much larger for you to be able to notice a difference whereas at lower frequencies it is slightly smaller.

You might expect the fact that we can detect a difference between 1 000 and 1 002 Hz, 1 002 and 1 004 Hz and so on, to mean that we can recognize or identify, say, ten different frequency sounds between 1 000 and 1 020 Hz, namely: 1 000, 1 002, 1 004, 1 006, etc., to 1 020 Hz. But in practice this is not the case, since the average person can only recognize about six different frequency sounds reliably. Even between 100 and 8 000 Hz it turns out that the ordinary person can again only *recognize and identify* six sounds, although he can tell the *difference* between hundreds of paired sounds within this range. This very low figure may surprise you but it seems to be a general feature of the way our brains work, for if we try the same thing with levels of loudness, judgements of saltiness, sizes of squares, lengths of lines, brightnesses, and so on, we find that people cannot normally name or recognize more than between four to seven different values of these variables. Furthermore, when we add to, say, a set of different frequency sounds, different intensities as well, the number of categories we recognize reliably does not go from 6 to 12, that is 6 for pitch and 6 for loudness, but only to about 8 or 9. An experiment which used six different acoustic variables: frequency, intensity, rate of interruption, on-time fraction, total duration, and spatial location, each one at five different levels so that there were 15 625 different sounds used, resulted in about 200 of these being reliably identified on average.

These figures apply to the average person and it is well known that some people have a very good ability to recognize different frequency sounds and can name any one of a large number of notes played on a piano for example. We do not know how they do this but in any case it seems to be limited to one sensory dimension and even if they are good at one task, they are no better than the rest of us at others.

There are many other fascinating aspects to this problem of how the brain uses sensory information and clearly it might be a considerable advantage if we could find ways to improve our abilities in this respect. But so far no really effective training techniques have been discovered although there is some evidence that experience in early life is significant in determining these abilities. One other feature of these judgements is worth discussing here: the brain does not make absolute judgements irrespective of experience, it judges things very much in relation to the context. You will almost certainly have noticed this when driving a car. When you first start out, 30 mph probably seems a 'neutral' kind of speed, 15–20 mph is felt to be 'slow' and 40–45 mph 'fast'. After driving for a while on a motorway at 70 mph, however, 30 mph is now judged as being 'very slow', 50 mph perhaps as 'slow', 70 as 'neutral' and 80–85 as 'fast'. This is one reason why people take exit curves or roundabouts off motorways too fast.

So the brain, generally speaking, judges sensory information in terms of five or six categories rather like 'very slow, slow, normal, fast, very fast' for speed, or 'very low, low, middle, high, very high' for sound frequencies. And the judged 'middle of range' value depends very much on the immediate past experience of the person doing the judging.

This has important implications for human information systems: unless we give a man numerical information (and even then the same effect can occur to some extent) he will judge magnitudes or intensities or other variables in terms of a very limited number of categories.

Section 5

Outputs from man

We have spent some time looking at input channels to man and it is convenient to compare these with the output channels. There are basically two forms of output—*physical effort* and *information output*. The physical output is generated by muscle activity and one of the characteristic features of man compared with other animals is his almost unique ability to exert physical force in an enormous range and variety of actions. Most of this variety comes from manipulation with our fingers and thumbs and it is the ability to oppose the fingers with the thumb which gives us the tremendous advantage of being able to grasp and manipulate so many differently shaped objects, especially objects which we can use to do things with, i.e. tools. The emergence of the first animals with an *opposable thumb* is regarded as a particularly important point in evolution.

physical effort and information output

The use of our muscles is controlled by specialized parts of the brain linked to the muscle groups in the arms, legs, trunk, neck, etc. The nerve fibres going to the muscles are called *efferent* fibres and the whole control system is usually called *motor control*. As its brain develops a child acquires the ability to control muscle contractions and relaxation in an organized way and gradually the brain is able to produce controlled movements and co-ordinated activities such as walking, running, jumping and so on. These output activities can be organized into very complex patterns which are learnt by experience; these are called *skills*.

motor control

skills

Not all motor activities are used to produce force as the primary objective or to move the body about, many are used to convey information. This is a vitally important function which in its most developed form results in talking and writing and the use of language. It is important to realize that these activities are the result of motor actions but that the exact motor mechanism is not usually directly relevant to the activity. If we want to convey a message, that is to communicate, we can speak it, write it, tap it out in morse code, or even sometimes simply adopt a particular body posture, as when we shrug. So for this reason, although communicating is done by using physical motor control of the body, we regard *communicating* as a separate psychological and social activity and we shall not discuss it further here as we have devoted a large part of the next unit to this important subject.

There are other forms of output from the brain which do not go directly to muscles but go to glands in the body. These are used to make adjustments to the body chemistry but sometimes they also convey information, as when we blush, or break out in a 'cold sweat'. These glandular or muscle-gland activities are usually indicators of emotions and contrast with the direct voluntary control of the skeletal muscles which are produced by the rational conscious part of the brain.

Decision-making and skilled behaviour

So far we have not said anything about the 'middle' function in our simple model of the brain—decision-making. We normally associate skills with complex and precise physical motor control of the body, as when we catch a ball or ride a bicycle. The basic idea of skills, however, can be extended to embrace the actions of the brain where the motor element is comparatively irrelevant as in controlling a chemical process plant—the knobs and switches have to be adjusted to the correct extent at the correct time but the switches could be moved by the foot or elbow, or even by instructing someone else. We shall be dealing with these *mental* skills in later courses. The role of decision-making will become clearer now in connection with skilled behaviour. When we do something *skilled*, we usually act in a particularly interesting co-ordinated way: we take in information via our sensory channels, we use this information in conjunction with stored or remembered information to make decisions, we then act out these decisions using the action or motor system. Often this action takes place incredibly quickly and at a subconscious level so that we are completely unaware it is going on. Watch a skilled knitter at work, the actions flow continuously with very precise control, only an occasional glance is needed to keep track of the wool, needles and stitches. Yet the behaviour requires a continuous series of decisions as to what precise movement is needed next and the movements are *monitored* all the time to ensure that they are appropriate and correct. One point worth noticing here is that the taking in of information is itself a part of a skill. The highly skilled man not only knows the right sequence of decisions and appropriate actions, he also knows *where to look* and *what to look for* (or feel or hear as the case may be).

The customary use of the term skill, then, is to describe the exercise of control over the muscles and, as you saw in the Systems unit, many skilled actions such as typing a sequence of letters like 'yours sincerely' can be done in an *open loop* manner. In practice, pure open loop control is rare and most skilled actions result from a combination of controlled muscle action co-ordinated with feedback information from various sources. The obvious major source of information is from the eyes, and to some extent the ears, but even without these sources to provide information we can still perform controlled movements because, as we have seen, we get feedback from touch sensors and from the proprioceptors within the body. This combination of sensory information with motor action is called *perceptual–motor behaviour* and the skills *perceptual–motor skills*. These have become sufficiently important branches of scientific study to have a scientific journal devoted solely to them, called *Perceptual–Motor Skills*.

perceptual–motor behaviour and skills

Clearly skilled behaviour is not simply a matter of co-ordinated sequences of, say, eye–hand co-ordination. Swinging a cricket bat in the prescribed and skilful manner for an off-drive is no use unless the bat is in the right place to strike the ball at the right time. This matter of *timing* the movements of the body to fit in with, or match, the movements of other objects is an essential part of skills as well. Take the batsman waiting to receive a ball from the bowler: the bowler starts his run-up and the batsman crouches in readiness. It is no good the batsman waiting until the ball reaches him to begin his stroke, there would be insufficient time for him to decide on the stroke, draw the bat back as he gets his body into position, and then execute the stroke. The ball would long since have passed him by.

timing

The batsman has to watch the bowler carefully and make a reasonable prediction of where the ball is going to go and what type of ball it is—is it going to pitch short or right up to him, is it swinging one way or the other, is it spinning? He must then prepare himself for a suitable type of stroke and make the necessary preparatory actions.

Example 2

The ball can leave the bowler's hand at anything up to 90 mph, the pitch is 22 yards long and the bowler releases the ball a yard past the wicket, with the batsman a yard from the other wicket. How long will the ball take to reach the batsman if it is bowled at 60 mph? Even the fastest of human reaction times under the most favourable circumstances is about 0·18 sec so there is only a short time available for the batsman to spot the ball and take the necessary action. If the batsman takes 0·2 sec to identify the type of ball bowled and a further 0·3 sec to move into position and prepare the stroke, where is the ball at the latest time the batsman can decide exactly what he is going to do? Work out the answer in metres from the bowler's wicket, assuming that the ball is released 1 m past the wicket and leaves the bowler's hand at 60 mph (96 km/h). How much time does the batsman have to observe the ball before he must decide what to do?

For answer, see end of unit.

Example 3

Now do **another** calculation for a service at tennis. The court is 78 ft long and you can assume that on service the ball travels at 80 mph. Where is the ball at the latest point where a receiver with a 0·2 sec decision time and a 0·2 sec movement time, standing on the base line, could still respond to it? (Assume that the ball travels straight down the court, not diagonally.)

How much time has the receiver available to observe the flight of the ball?

For answer, see end of unit.

From these examples you can begin to see how the information flow within the body to and from the brain can be an important factor in skilled performance.

Now try an experiment on decision-making for yourself. You will need a pack of cards and a stop-watch or a watch with a good seconds hand. You will also need a helper to time your actions. The idea is to see how long it takes you to make a series of decisions. Since it is difficult to time accurately the rather short intervals of a single decision or decision and action, take the time to do, say, thirty actions. If you then divide the result by 30 you will get the time for one action.

First shuffle the cards and then simply deal them into three piles by turning each card face up to you and placing it at random on one of the piles. Get the time for thirty cards and divide by 30 to get the basic *movement time*. Now collect the cards, shuffle them, and time the dealing process again, only this time sort the cards into two piles, black cards, and red cards. You can simply take the time to sort the whole 52 cards and divide by 52. Now fill in the result in the table and then repeat the process for sorting the 52 cards into four piles of the four suits. Repeat again for eight piles, using the four suits of non-court cards, and the four suits of court cards, so that you have eight different categories. Now take out the court cards and repeat the sorting with ten piles of cards using the ten numbers ace to ten. When you have completed the table, do the experiment again on your helper. Now make a graph of the results by plotting the time taken *per sort* against the number of choices which had to be made. Subtract the movement time which you determined in the first trial from the sorting time so that you are left with only the time taken to recognize and decide which pile to put the card in. We can break down the activity as shown in Fig. 23.

no. of piles	time per sort		mean time
	1st trial	2nd trial	
2			
4			
8			
10			

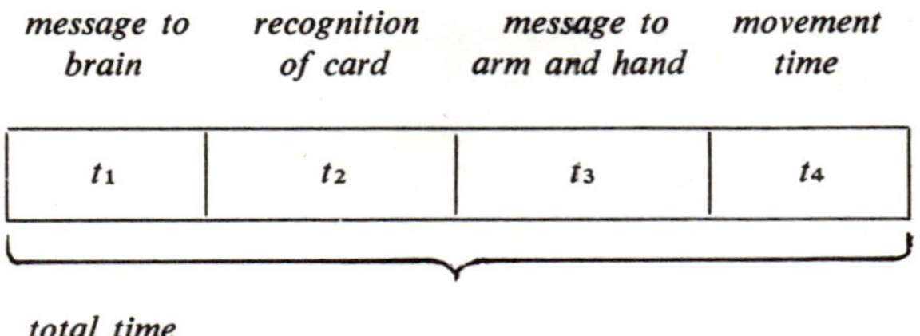

Fig. 23.

This shows that the total time is made up of various parts, some of which you have measured and some which you can calculate from known facts.

However, the important point is not so much the nature of the overall time but the way the time changes as the task changes. In your experiment the results will probably look like the graph in Fig. 24 taken from an experiment which one of us did.

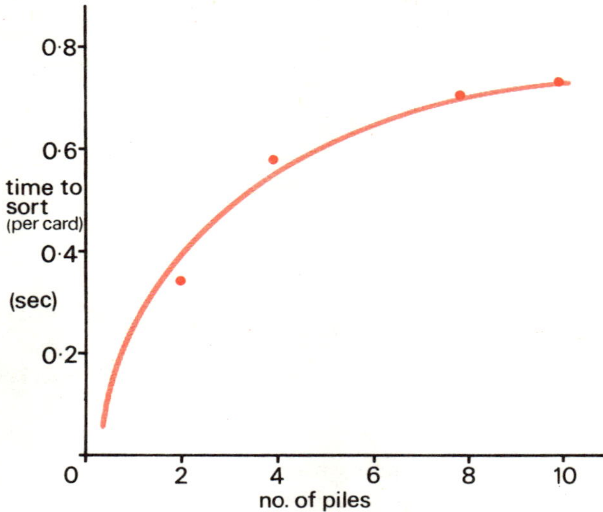

Fig. 24 *This graph shows the results obtained when sorting cards into piles of 2, 4, 8 and 10 different categories. We allowed a little time for the subject to practise dealing and sorting so that he used the same method all the time.*

This is an interesting curve and we want you to re-plot your results, but this time using the logarithm* of the number of choices which had to be made. The values for the logarithm (to base two) of the number of choices are given in Table 2.

The result is perhaps more interesting now because if your results are like ours you will find that the points fall roughly on a straight line. Straight line relationships are always of interest in scientific investigations because they suggest that there is a simple rule relating the quantities concerned to each other. In this case we can express the connection as:

$$T = k \log_2 n$$

where T is the choice time
and n is the number of choices
and k is a constant.

From our earlier discussion it looks as if it is the *decision time* which changes as we have to make more choice decisions. As you might expect in any study on human beings, things are not quite as simple as this. With a great deal of practice (about 10 000 trials in fact) we find that people can make choices among eight things as quickly as they can between two.

Table 2

n = number of piles or choices	$\log_2 n$
2	1·00
4	2·00
8	3·00
10	3·32

decision time

* If you are unfamiliar with logarithms, refer to Modelling I.

Further, if the stimulus and response are naturally related, as when you have to push the upper button on a lift to go up and a lower button to go down, then again there may not be much more time taken to choose from a larger array of choices. Incidentally, this idea of the 'naturalness' of a response to a signal is important and it is called *compatibility*. A highly compatible response is one where a majority of people in a given population accept that the response required is a 'natural' one. Note that this can differ for different cultures—in Great Britain we expect lights to go on when we push a light switch down, in North America they expect the opposite. Another example is the difficulty often experienced in steering a boat with a tiller after being used to cars. The tiller has to be pushed the opposite way to that in which you want to go, compared with a car where turning a steering wheel clockwise to go to the right is a 'natural' connection, i.e. highly compatible, but with some radio sets, for example, the tuning knob works the opposite way.

The way the brain makes decisions of this kind is still not completely understood and there is a great deal of research going on into these topics. Current thinking is that the brain assembles evidence or information and makes a decision on probabilities or likelihoods of outside events having in fact occurred. This can be shown by looking at other factors which affect the situation, for example the frequency of occurrence of the different choices, the ease or difficulty of discriminating among the choices and, of course, practice. In our card-sorting experiment the different choices did not occur with equal frequency (there are more number cards than court cards) and it is clearly harder to distinguish between say a king and a queen than, say, an ace and a ten.

Complex skills

So far we have talked about simple sequences of *sensing—deciding—acting* behaviour. These are more commonly called *information—decision—action* or IDA sequences. More complex skills are built up from patterns of such sequences but we should first look more closely at what is implied in these IDA sequences. Although the act of turning a playing card face upwards, recognizing that it is, say, the ten of diamonds, and placing it on a certain pile, seems very simple, it is in fact a very complicated information processing task. We know very little about the way the brain 'sees' the card, and how it recognizes that it is a playing card and a particular one at that. The action aspect is less complex because we can more readily understand how a simple rule such as 'place diamonds on the left-hand pile' can be operated once the card has been recognized. The brain carries out a large number of processes of 'recognition' and 'understanding' of objects or situations about which we have no knowledge at all. We certainly cannot build any kind of device which can recognize such a variety of different things, nor at the moment can we conceive a way in which this might be done.

IDA sequences

But these simple IDA sequences are not sufficient in themselves to explain skilled behaviour because one of the most significant things about skills is the *timing* aspect, as we have already said, and this implies that the brain is not just taking in information about what is happening and making the appropriate response, it is also *predicting* ahead what will happen. When we make a shot in tennis we move the racquet *before* the ball arrives and arrange the movement so that the racquet meets the ball at the right point in space. This requires us to predict the *future* position of the ball. All physical objects have inertia and when we try to control their motions we must take this into account. Stopping a car is not just a matter of putting on the brakes—the position of the car at rest is some way past the point where we apply the brakes and we must know this if we are to stop successfully at a traffic light for example. This may not seem very difficult where the distances are relatively short but it becomes a problem with a high-speed train or a large tanker which may take several miles to stop. You are clearly in trouble if the point at which you can first stop is far beyond your vision! A typical example of this does in fact occur with cars travelling behind each other on a motorway.

Example 4

Suppose ten cars are travelling comfortably at 60 mph each 50 ft behind each other. Suppose it is night time and each driver can only see the brake lights of the car in front. Now suppose that the first driver sees a blockage some way ahead, say 260 ft away. He applies his brakes and stops fairly comfortably 260 ft later (the highway code gives the shortest braking distance at this speed as 180 ft). If all the following drivers have the same reaction time from seeing the brake lights of the car in front go on to operating their own brakes, which the code gives as about 0·68 seconds (or 60 ft at this speed), what happens to the cars if they try to pull up just behind the car in front? (We shall ignore the length of the car.) The second driver sees the brake lights of the first go on (when he, No. 2, is 310 ft from the obstruction); he is 50 ft behind the first car and he travels 60 ft while reacting, he then brakes within the remaining 250 ft, again fairly comfortably. The third driver sees the second's brake lights go on (No. 3 is then 300 ft from the obstruction). No. 3 travels 60 ft while reacting and then brakes himself, stopping within the now available 240 ft. Now carry this calculation on further, where do drivers No. 7

and 8 stop? And Nos. 9 and 10? What is the effect, if any, of including the length of the car, assuming each is 10 ft long and the separation distance is 50 ft. from *front bumper* to *front bumper*?

Here an accident can only be avoided if the drivers anticipate the stopping of the car in front by watching the brake lights of the car before that. Will this avoid the accident no matter how many cars there are in the convoy? If not, how many cars can travel safely at this speed and separation distance before an accident is inevitable (assuming that each driver responds to the brake lights of the car two in front of him)?

For answer, see end of unit.

Clearly the problem arises because the cars are too close together and larger separation distances are necessary. In fact, the problem is more likely to arise at high speeds because, as you can see from Table 3, the braking distance increases more with speed than does the thinking reaction-time distance.

Table 3

Speed (mph)	Reaction-time distance (ft)	Braking distance (ft)
20	20	20
30	30	45
40	40	80
50	50	125
60	60	180
70	70	245

Anticipation then, plays a part in a skill but, of course, there are many other factors. We must know *what* to do as well as *when* to do it. When we first learn to drive a car we have to learn what actions are necessary to start the car and to control it. We then practise these actions and build up learned patterns of co-ordinated actions such as going through a gear-changing sequence. We then have to learn when to apply these operations and how to integrate them to produce the desired output from the car—driver system. A complete analysis of the skill of driving a car is still waiting to be done and, although a good deal is known about what constitutes *good* driving so far as the highway code is concerned, we do not know so much about *good control* procedures such as the best way to deal with a skid.

7.1 Getting information

Complex skills are composed of a number of stages or sub-skills which correspond to the IDA sequences we have mentioned. First the controller or operator has to have the appropriate *information* about the situation. This entails two things, *obtaining* the necessary information and then *organizing* it for further treatment. Obtaining information is a sub-skill in itself—poor drivers are known to look at their rear-view mirrors either too infrequently or at the wrong moment. Man has a vast number of sensory information channels but for high-level conscious control activity he can only attend to a very limited number of things at a time. We can normally only look at one dial or instrument at a time and if, for example, we want to know both what is in front of us and behind us in a car, we have to switch attention from front to back to front again in some ordered fashion. This switching from one information source to another is called *sampling* and there is good evidence to suggest that skilled performers have acquired very appropriate and efficient *sampling strategies* to extract the necessary information from the environment around them.

> Try listing the information sources used by a car driver as he approaches a roundabout and takes one of the far exit roads.

Knowing where to look or which signal or information source to attend to is one thing—how often you need to sample is another. In the Systems unit we suggested that a continuous system could be treated as a discrete one by taking suitably frequent measurements of the variables, which is something like this sampling process; in fact, it can be shown that there is a certain frequency of observing a changing variable which will enable you to follow all that it does even though you are not watching it continuously.

But having got the information, what do we make of it? We have probably all had the experience of looking vaguely at a complex object while a skilled person points out the significant features. A doctor carrying out a diagnosis looks at and feels the patient as well as extracting information by questioning, and then organizes all his data into a meaningful pattern. Of course, this is a continuous process: we first notice one feature and eliminate some possibilities, then look for a second feature, and so on. The skilled man somehow seems, nevertheless, to see *more* of what is there than we do. He notices significant features and *organizes* the information available into a *recognition* that some particular fault or situation is occurring.

7.2 Using information

With the organized patterns or pieces of information available, the *decision mechanism* of the brain can start to work. Comparisons are made with past experiences, factual information is retrieved from memory, analyses, predictions or probabilities are worked out, and the brain arrives at decisions as to what could be done. Moral and ethical factors are considered, of course, and our emotional state will also influence our thinking and deciding. Experiments have shown conclusively that even the information which arrives at the decision centre is distorted in some way and that our emotional states and our past experience from the culture we have grown up in can all affect the information assessment and decision processes. For example, the famous illusion, the Müller-Lyer, shown in Fig. 25 is quite effective in the modern Western world, but people in relatively primitive *non-rectangular* cultures apparently do not 'see' the illusion in quite the same way.

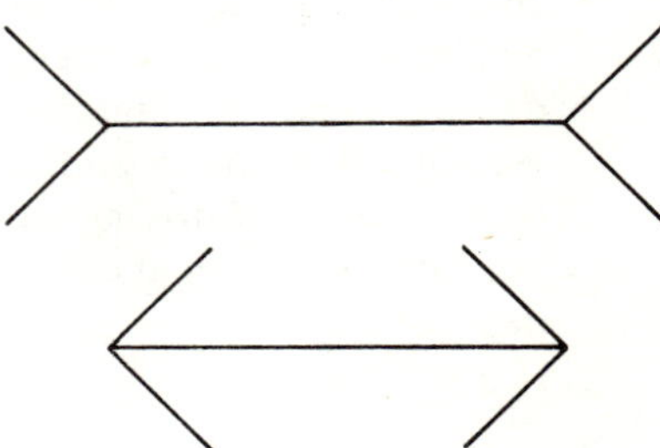

Fig. 25 *The lines between the arrow heads are equal in length.*

7.3 Rule following

Much of our skilled behaviour is governed by the fairly straightforward application of *rules* and we can identify these from the way people behave. If we are tuning a radio to a new station, the rules might look like Fig. 26.

> Write out for yourself a similar sequence for starting a car and driving it off from the side of the road.

You can see from these examples how the decisions fit in with the information and actions necessary. There is, of course, much more to decision-

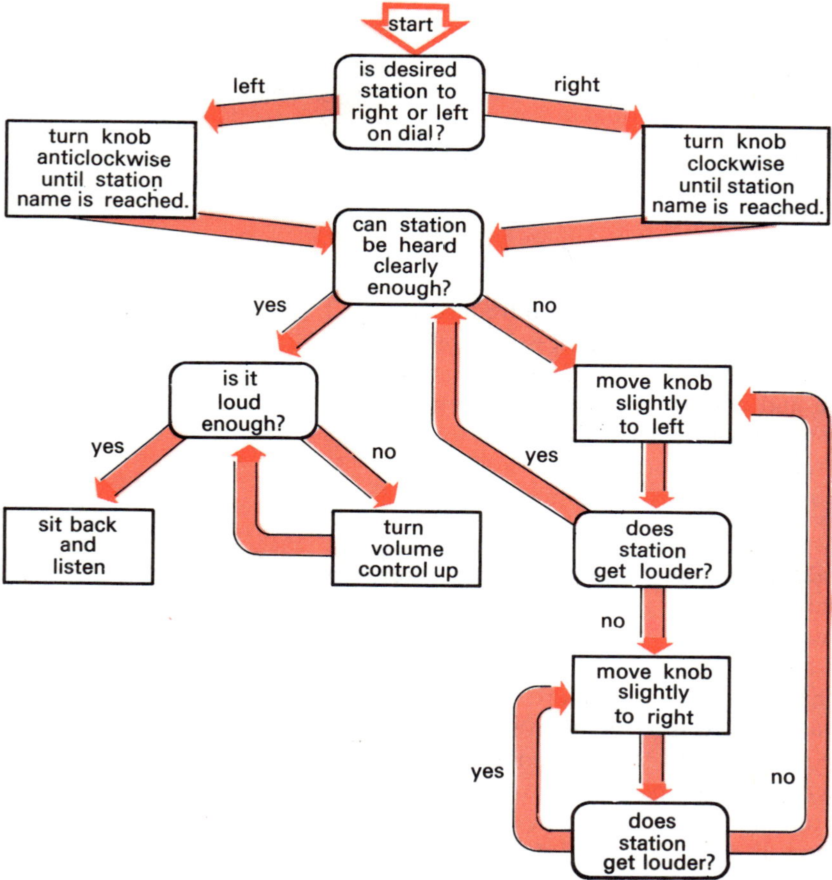

Fig. 26.

making than just applying rules—the rules have to be built up in the first place. Some people may have discovered or worked out better rules than others. The time available to do the task may be so short that we cannot apply all the rules we in fact know, or vital information may have to be guessed at with varying degrees of success (management decisions are often of this type).

7.4 Taking action

Once a decision has been made we take action and here again there is a good deal of learning needed. We all experience situations where we know what to do, and when to do it, but somehow we just cannot organize our fingers or body into the correct positions or movements. This acquisition of movement and action sequences takes time and in general has to be learnt by the often painful process of trial and error. A good deal is known about the best way to train people in motor movements although it is possible that we still do not exercise all the control over our muscles that we might if we only knew how.* Compared with the information input and decision processes, motor actions are relatively accessible and with modern aids, such as slow-motion filming and magnification of movements, we can learn a great deal about actions.

* Some dancers in the East can control individual muscles all over the body at will.

The man-machine system

Our study of the human component has been designed to introduce you to some of the properties of man and his information-processing abilities. The last part of this unit takes a brief look at how man and machine can be 'married' to make an effective and efficient working unit. A man–machine system is any combination of man (or men) with some form of tool or equipment. A man with a spade is a simple form of such a system and one of the most complex is probably a spaceship capsule used to send men to the moon.

As we have noted earlier, machines have mainly replaced human effort and the man part of a man–machine system is primarily exerting a controlling or directing function. To do this the man must be able to get the requisite information needed and also be able to exercise control. With simple and primitive equipment or tools man still has more or less direct access to the information sources and he acts directly on levers or simple mechanical devices to control the system. With more advanced systems, man gradually loses this direct and relatively simple linkage with the world out there and he has to rely more and more on information coming from artificial sensors via instruments and displays. Air speed, height, rate of climb, outside temperature and so on, all have to be sensed and relayed to the pilot of a modern aircraft. A process operator receives information about flow rates, pressures, temperatures, reaction rates and so on from a chemical plant. The contact between man and machine becomes highly selective and *artificial* in the sense that the pilot or process operator does not see or touch or hear the real things, but only gets instrument readings. We refer to all such information sources as *displays* and the selection and design of these displays is one factor of vital importance in determining just how well a man can cope in the system.

displays

Controls are likewise now dissociated from the actual effective moving parts. Servo-mechanisms operate ships' rudders or car brakes so that a slight touch on a small wheel or pedal is sufficient to move large control devices. Controls (in the sense of the knobs and levers used by man) and displays are usually grouped together and this array of information sources and action mechanisms is presented as the *interface* between man and machine.

interface

The gap or interface barrier which has emerged between man and the machine and the outside world may be dangerous as well as beneficial. Perhaps it would be more accurate to talk of the limitations imposed by this artificial interface, because even today interface displays and controls tend to give us the *easy* things to present rather than the *necessary* things. Driving or flying 'by the seat of one's pants' is normally taken to refer to the intuitive aspects of skilled performance but there can be little doubt that information on the direction of motion and acceleration of a vehicle was and is obtained via the slipping of pants on the seat! Now that we are all strapped into contoured seats we no longer get what might be useful information of this kind. Power steering makes operation of the steering wheel easier, but the driver can also lose the 'feel' for the kind of adhesion he is getting between the tyres and the road. The difficulty arises from the fact that we do not know enough about the use of all the many different sensory channels by man and when we cut out the use of some (for good reasons, of course) we do not realize the problems we may create. It is only when we come to build, or try to build, automatic control devices that we realize just how much information is needed for effective control. Automatic pilots and landing equipment are very impressive, but the use of these devices means that a great deal of effort is put into the supply of appropriate information to them. If some of this effort had been devoted to giving pilots the information they need in a form which they could readily use, the human controller's performance might have been even more impressive. Of course, in many cases machine control devices are inherently 'better' than human ones—they rarely make mistakes, they do not get tired and they go on working 24 hours a day. But there is often a serious sacrifice of *flexibility*; the machine cannot cope with unusual or unforeseen emergencies whereas a human being can. What we need is a systems approach to look at the whole man–machine–environment system.

There is much more to be said on this topic but it will have to wait for later courses. For the moment it is worth noting that the important thing is to make the most effective use we can of man in combination with machine. Man has his own unique and specialized capabilities, machines have theirs; we have hardly begun to explore the possibilities of the marriage between man and machine yet.

Suggestions for further reading

Alpern, M., Lawrence M. and Wolsk, D. (1967) *Sensory processes.* Wadsworth Publishing Co.: Prentice-Hall.

Fitts, P. M. and Posner, M. I. (1967) *Human performance.* Wadsworth Publishing Co.: Prentice-Hall.

McCormick, E. J. (1964) *Human factors engineering.* McGraw-Hill.

Weintraub, D. J. and Walker, E. L. (1967) *Perception.* Wadsworth Publishing Co.: Prentice-Hall.

Welford, A. T. (1968) *Fundamentals of skill.* Methuen.

ACKNOWLEDGEMENTS

Acknowledgement is made for the following material used in this unit:

Figures 8, 9 and 10: G. S. Holister, *Experimental Stress Analysis*, Cambridge University Press; Figure 19: D. Henry Edel jr., *Introduction to Creative Design*, Prentice-Hall Inc., Table 3: *The Highway Code*, Department of the Environment and HMSO.

Answers to examples

Example 1 (Section 3.4)

Because it can act as a general warning or alerting system for events or dangers which we are not attending to at the time.

Example 2 (Section 6)

The ball will travel the 20 yards in 0·68 sec. The ball will travel 44 ft in the 0·5 sec (0·2 + 0·3 sec) which the batsman will use up in deciding on his stroke and in moving into position. So the last point for decision is 44 + 3 = 47 ft from the batsman's wicket which thus is 5·8 m from the bowler's wicket. Since the bowler delivers the ball about 1 m past the wicket, the batsman must decide what to do on the basis of 4·8 m of travel of the ball after it leaves the bowler's hand. This takes only 0·18 sec! The batsman does, of course, have some help from his knowledge about the bowler, whether for example he is known as a leg-break bowler, and from the way the bowler runs up to the wicket—and also from the immediate past performance of the bowler.

Example 3 (Section 6)

The ball would travel 46·9 ft in the time the receiver decides and acts so the ball would be 78 − 46·9 = 31·1 ft away from the server at the latest time the receiver must decide what to do. The ball takes 0·26 sec to travel this distance which is therefore the time available to the receiver. Again, the receiver is assisted by the knowledge he has of his opponent's past behaviour.

Example 4 (Section 7)

Clearly cars 7 and 8 can pull up before the obstruction since they start braking at 200 ft and 190 ft respectively from the obstruction. Car 9 just manages to stop, since he brakes at 180 ft from the obstruction and can just stop in this distance. Poor car 10 has only 170 ft available and therefore must crash into the obstruction or the rear of car 9.

If we assume each car is 10 ft long, then the cars still start braking at the same points as before but they now have less distance in which to stop because there is a line of stationary cars, each 10 ft long, building up at the obstruction. In these circumstances car 6 will hit car 5.

If the drivers watch the car two in front then they always have enough time or distance to stop in because the warning signal always occurs more than 60 ft away from them.

Fig. 12 *Standard eye chart reduced to one-quarter actual size.*

The Open University

Technology Foundation Course Unit 3

The week number during which this unit should be studied is not necessarily the same as the unit number. Please consult your wall-chart study guide to the Technology Foundation Course to find the place of this unit in the course.

SPEECH, COMMUNICATION AND CODING

*Prepared by John Sparkes
for the Technology Foundation Course Team*

THE OPEN UNIVERSITY PRESS

Aims

The main aim is to explain, through a study of the properties of speech and other signals, some of the characteristics of oscillations, signals and telecommunication systems, and to suggest possible strategies for the design of error-free engineering systems in general.

Subsidiary aims are:

To describe sinusoidal vibrations and oscillations and show how they can be produced. To explain resonance, spectra and elementary ideas of Fourier analysis of periodic waves.

To describe speech signals in detail so that their complexity can be appreciated and to compare them with telecommunication signals.

To develop the ideas of information and redundancy and to show that these can be, to some degree, measured with practical signals.

To show how redundancy can be added to simple signals to achieve simple error correction.

Objectives

After working through this unit you should be able to:

1. Define the following terms and explain the underlying ideas: frequency, resonant frequency, damping, forced vibration, sound wave, wavelength, wave velocity, waveshape, periodic wave, aperiodic wave, Fourier component, frequency spectrum, fundamental, harmonic, phase, amplitude, formant, carrier, modulation, demodulation, bandwidth, phoneme, redundancy, information, bit, error-correcting code.

2. Describe what is meant by a sound wave, how it is generated and how it may be analysed into its constituent frequencies.

3. Describe the mechanism by which speech is formed and the main characteristics of the resulting sound waves.

4. Discuss some of the problems which must be solved to achieve full understanding of how a person can recognize and interpret speech.

5. Compare and contrast the salient characteristics of speech with those of machine-generated signals.

6. Explain what is meant by redundancy in a signal and how it can be used to offset the effects of errors in the signal.

7. Explain what is meant by 'information' in the technological sense and carry out simple calculations on the information content of various types of signal.

8. Explain parity check code, block code and chain code methods of error detection and correction.

9. Discuss the general principle of introducing redundancy into technological systems and be able to quote some examples of how and why this has been done.

What you have to do

There are five simple exercises in Section 2, two in Section 3, one in Section 6 and two in Section 7. Also two problems in Section 2.3 and two problems in Section 7.3 serve as self-assessment questions. A simple experiment in Section 5 will require the help of another person.

For revision, check yourself against the objectives and go over the assignment questions.

Note to student

Where references are made in Course Units to the T100 Book of the Course, *The man-made world*, we are using the following abbreviations:

T100/BK (Intro)	Introduction to *The man-made world.*	
T100/BK1	Case Study 1.	The electricity supply industry.
T100/BK2	Case Study 2.	Education by satellite in Brazil.
T100/BK3	Case Study 3.	The plastics and steel industries.
T100/BK4	Case Study 4.	Transport.

Contents

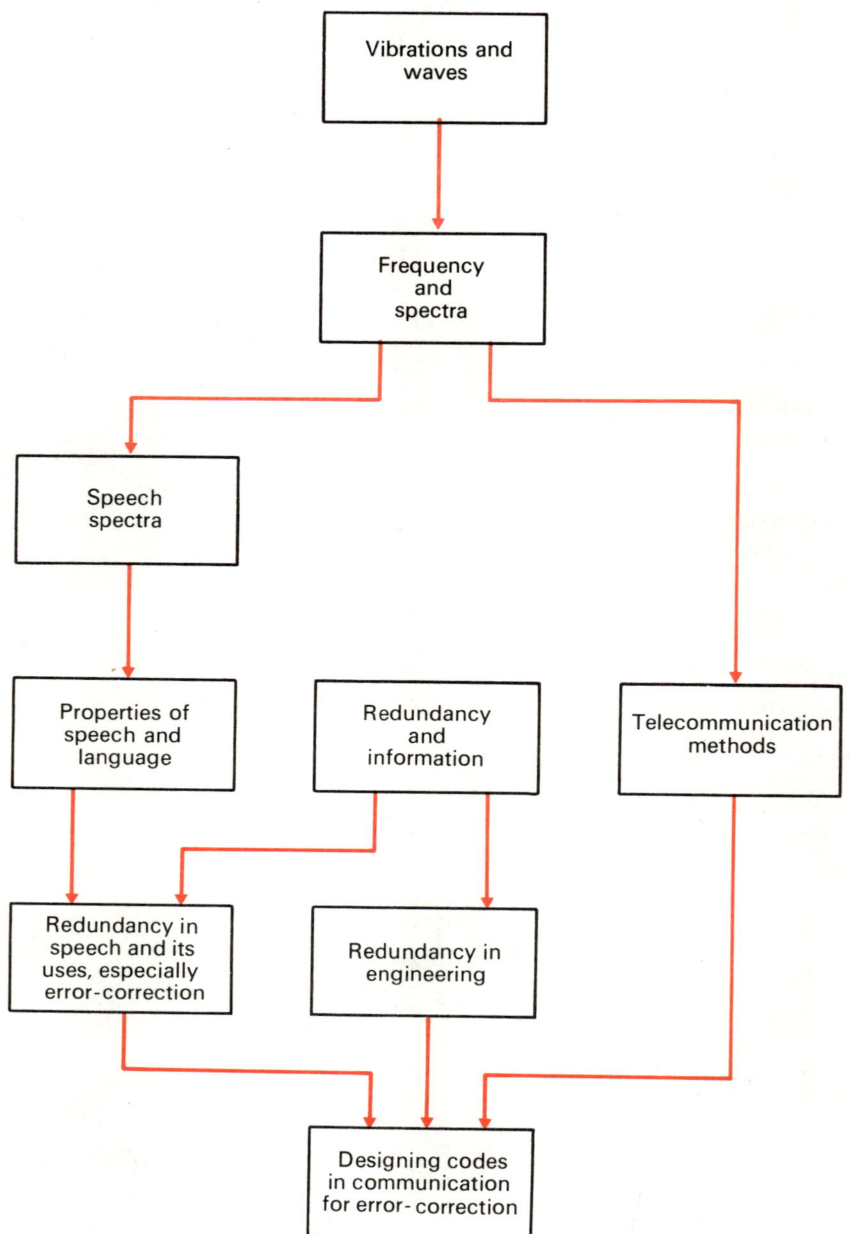

Vibrations and waves
Frequency and spectra
Speech spectra
Properties of speech and language
Redundancy and information
Telecommunication methods
Redundancy in speech and its uses, especially error-correction
Redundancy in engineering
Designing codes in communication for error-correction

Introduction

The behaviour and organization of systems depends crucially, as we have seen in the previous units, upon the communication between various parts of the system. Human beings, and the society in which human beings live, are systems for which this is especially true. Yet, for some reason, speech, the basic form of communication between us, and which we all use every day of our lives, is something to which little attention has been paid until very recently.

Much of technology is concerned with providing ways of conveying spoken communication between human beings by telephone, radio or television but we have done little to find out what speech is really like. We just use it.

Somehow through constant use as a tool essential to our daily lives, speech has become an extremely efficient system for the exchange of ideas. It is not always easy to communicate the most complex ideas and no doubt many arguments arise through misunderstandings in the use we make of speech. Nevertheless, it is broadly true that if any one of us has an idea he wants to express he somehow knows how to convert this idea into language and to articulate the sounds which constitute the appropriate sentences. The rest of us then somehow know how to interpret these sounds and apprehend the idea that initiated the activity in the speaker's mind. Speech has become capable of performing this function despite the many different voices, speaking habits, accents and dialects of those who speak the same language, often in conditions of severe noise.

But why should a study of speech form part of a course on technology? There are three quite distinct reasons. First, communication between people is at the heart of human community-living and the business of assisting communication between people by using technology is one of the major activities of advanced industrial societies.

Second, the way speech is produced by the vocal organs illustrates many of the general properties of vibrations as well as of the techniques used in telephones, radio and other kinds of telecommunication.

Third, speech has a quite distinct property, which we are only just beginning to understand, of being comprehensible despite severe noise, distortion and variation. A signal of this kind is so different from the kinds of message-bearing signals used by technology that it seems that technologists have much to learn from it.

In the previous unit the idea of the frequency of audible signals, of notes in music for example, was introduced without much explanation, because it was assumed, I hope rightly, that most of you would know pretty well what the words 'a frequency of 500 cycles per second' mean. In this unit we need a more precise understanding of this and related concepts (resonance, bandwidth, spectrum, etc.) whether we are discussing speech or telecommunication or, later in the course, noise and mechanical vibrations. So to begin with we will discuss the background of these ideas, through a consideration of vibrating systems and vibrations in general.

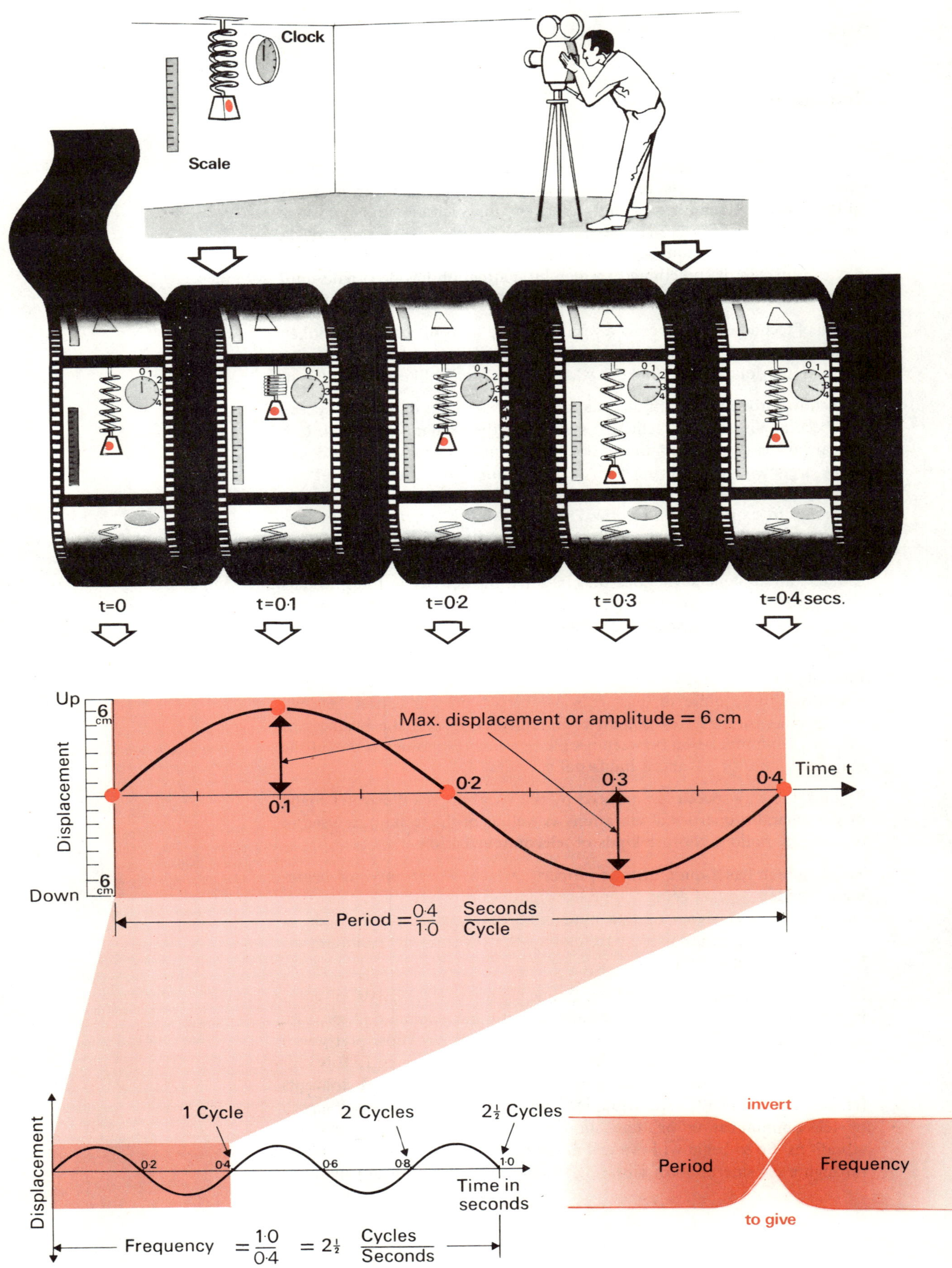

Figure 1   *Displacement, amplitude, period and frequency of an oscillating mass.*

Section 2

Vibrations

Perhaps the best way to approach the subject of vibration is in terms of a simple example. There are many to choose from, such as the vibrating prongs of a tuning fork, an oscillating piano string, a pendulum or a spring and mass.

Fig. 1 shows in graphic form the behaviour of a mass vibrating up and down on the end of a spring. From this figure the ideas of the *amplitude*, *frequency* and *period* of a vibration should become clear to you, if they were not already familiar ideas. It also shows how movement can be drawn on a graph. Such a graph can be used to describe the behaviour of most common vibrations, not just masses hanging on springs, but pendulums, tuning forks and so on.

A common factor in all examples of a vibrating system is the existence of a restoring force which acts on the body when it is displaced.

What are the restoring forces acting on a child swinging, a balance wheel of a watch, a springboard?

Gravity, the tension of the hair spring and the elasticity of the diving board.

Another common factor is that all vibrating systems have a characteristic frequency called the *resonant* frequency. Disturb them and they will vibrate at their resonant frequency—whatever the amplitude, more or less; the form of the vibration tends to be *sinusoidal*, like the curve of Fig. 2a.

A third characteristic is that the vibrations always die away owing to friction (in the air, in the bearings or in the material from which the system is made). This is called *damping*. The three vibrations plotted graphically

vibrations

amplitude
frequency
period

resonance

sinusoidal wave

damping

(a)
(b)
(c)

0 1 2 3 4 5 6

Time in seconds

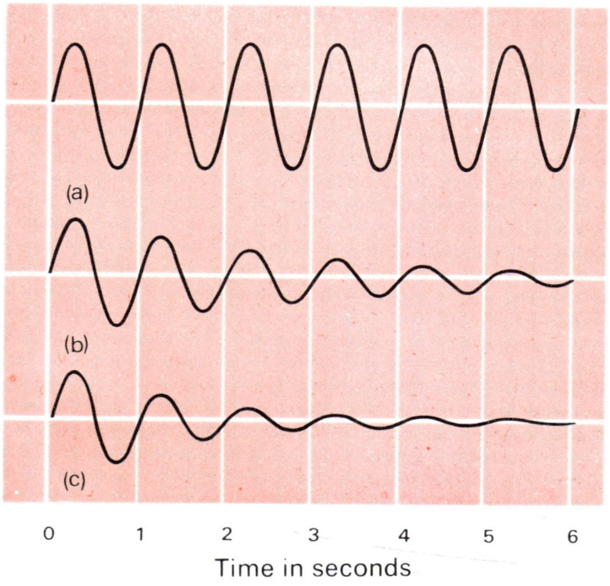

Figure 2 Displacement of the vibratory mass with and without damping: (a) no loss; (b) lightly damped; (c) more heavily damped.

in Fig. 2 show varying degrees of damping. The first is a pure undamped sinusoidal vibration and never quite exists in nature, the second and third are successively more heavily damped.

What are the frequency and period of the first wave shown?

Now when we try to *force* a vibrating system into vibration at frequencies not necessarily equal to its resonant frequency interesting things happen. Consider a simple pendulum (a button on a piece of cotton about 10 cm long can be used to check our considerations). If, instead of pushing the weight at the end, we move the support back and forth, we find that the weight does not faithfully follow the movements of the support. In particular, if the support is moved only a little back and forth at the resonant frequency of the pendulum, the *amplitude* of the pendulum will steadily build up till it is much bigger than the amplitude of the movements of the support. This is what is meant by resonance. Movements of the support somewhat slower or faster than the resonant frequency can produce resonant response from a resonator, but of lower amplitude. Indeed it is possible, as shown in Fig. 3, to plot the response of a resonant system to excitation of different frequencies. Good resonators—those which would vibrate freely themselves for a long time—show very sharp response curves as shown by the black curve. The red curve corresponds to a more heavily damped resonator.

forced vibration

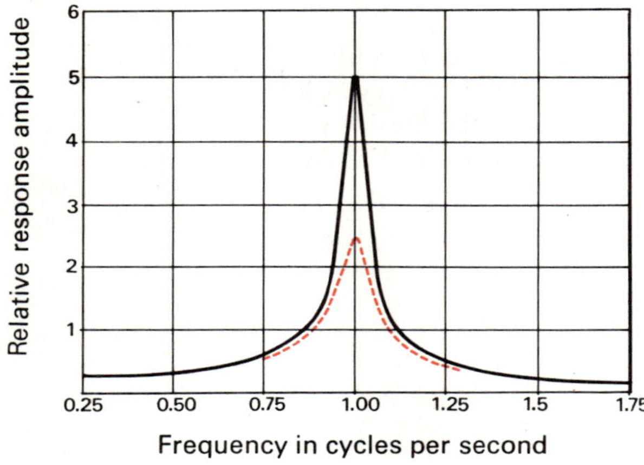

Figure 3   A frequency response curve with a resonant frequency of 1 Hz.

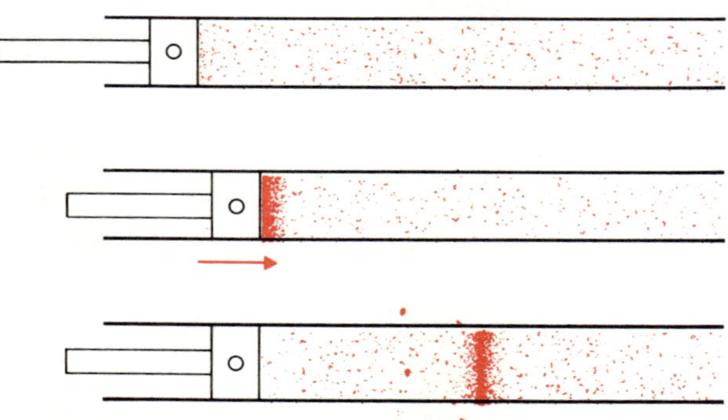

Figure 4   The effect of piston movement on air molecules in a tube. The zone of compression produced by the piston moves along the tube at the speed of sound.

2.1 Sound waves in air

All objects on Earth are surrounded by air. Air consists of many small particles, more than 400 billion billion in every cm³. These particles move about rapidly in random directions. We can explain the generation and propagation of most sound waves without considering these random motions as such. It is sufficient to assume that each particle has some average 'stable' position from which it is displaced by the passage of a sound wave. Each particle experiences an average force resulting from the continued bombardment of neighbouring molecules, which acts on the particle just like the restoring force in any vibrating system. If one particle is disturbed—moved nearer to some of the others—this force tends to push it back to its original position. Thus, when air is compressed, pushing the particles closer, this force tends to push them apart. By the same token, when air particles are separated by more than the usual distance, a force develops that tends to push them back into the emptier rarefied space.

This inherent 'springiness' or elasticity of air allows for the efficient transmission of sound in the form of pressure waves (Figs. 4, 5 and 6).

pressure waves

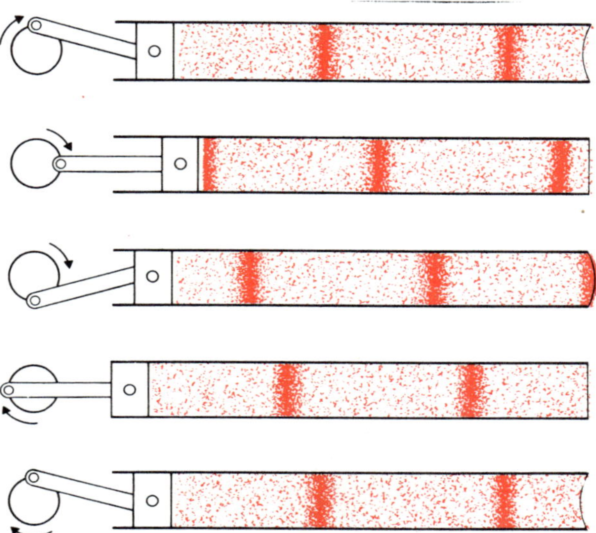

Figure 5   The effect of repeated piston movements on air molecules in a tube. The wavelength is easily seen as the distance between regions of highest compression.

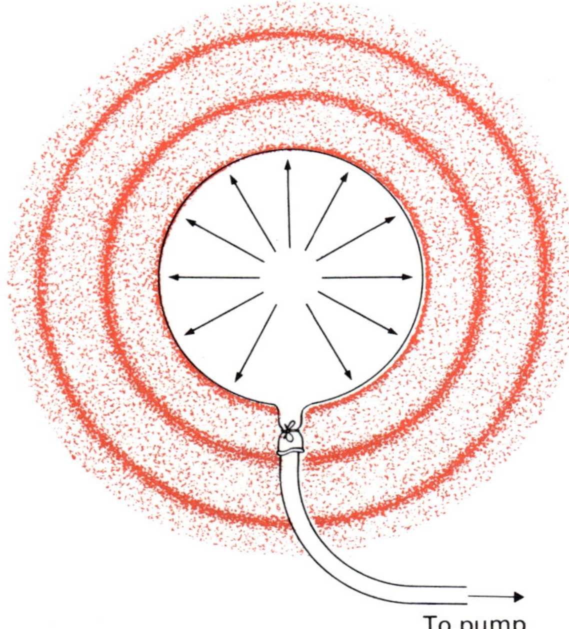

Figure 6   Sound waves produced by a pulsating balloon.

We see that only the disturbance itself (the vibration) moves through the air particles, which themselves simply move back and forth about their 'rest' positions. A *sound wave* is the movement (propagation) of a disturbance through a material medium such as air, *without permanent displacement of the particles themselves.*

Surface waves on water exhibit some of the characteristic features of sound waves. Water waves are vibrations of water particles much as sound waves are vibrations of air particles. The chief difference between the two is that, in sound waves, the air particles vibrate in the direction of wave movement, while, in surface waves, the water particles principally move up and down at right-angles to the direction of wave movement, though they move back and forth too. Instead of the compressions and rarefactions peculiar to sound waves, water waves appear as crests and troughs on the surface of the water (Fig. 7).

The distance between two successive compressions (or between two water wave crests) is called one *wavelength*. A wavelength is also the distance the wave travels in one cycle of vibration of the air particles. If there are f cycles in one second, the wave will travel a distance of f wavelengths in one second. Since the distance travelled in one second is the velocity, it follows that the *velocity is equal to the product of the frequency and the wavelength.* Since the velocity of sound is about 320 m/sec the wavelength of a sound wave whose frequency is 20 Hz is about 16 m. If the frequency is increased to 1 000 Hz, the wavelength is shorter. At 20 000 Hz, the wavelength is about 1·6 cm.

The average motion of the air particles is an important characteristic feature of a particular sound wave. We can plot the displacement of a particle from its rest position, instant by instant. The form of such a curve is called the *waveshape* and is not normally sinusoidal (as in Fig. 1): it could be random or pulse-like for example. In measuring sound waves it is usually convenient to measure and plot the sound pressure variations associated with the wave, and not the particle displacement itself.

sound wave

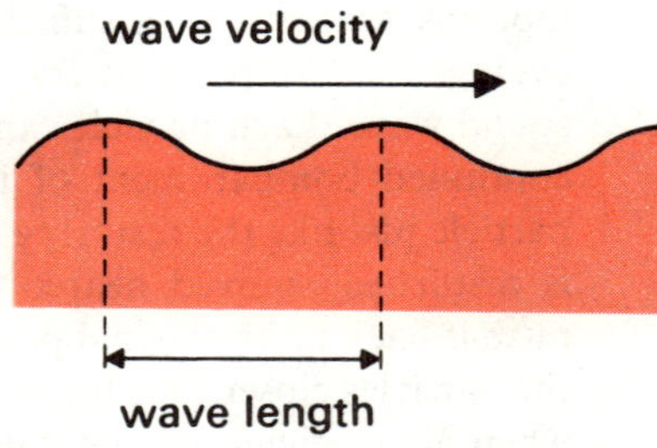

Figure 7 Crests and troughs of a water wave, showing direction of movement and wavelength.

wavelength

waveshape

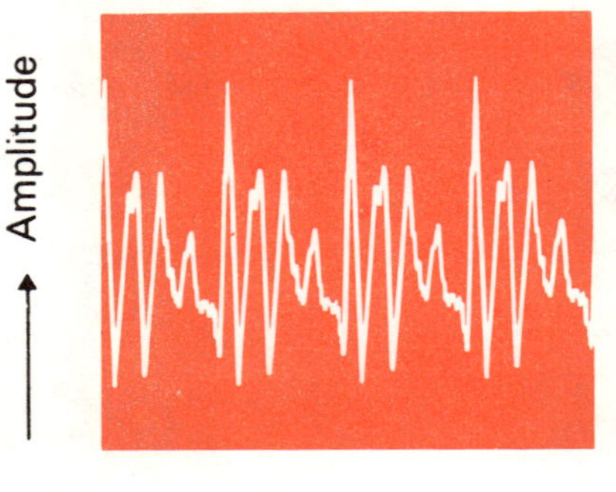

Figure 8a A periodic wave (a typical waveshape of the speech sound 'ah').

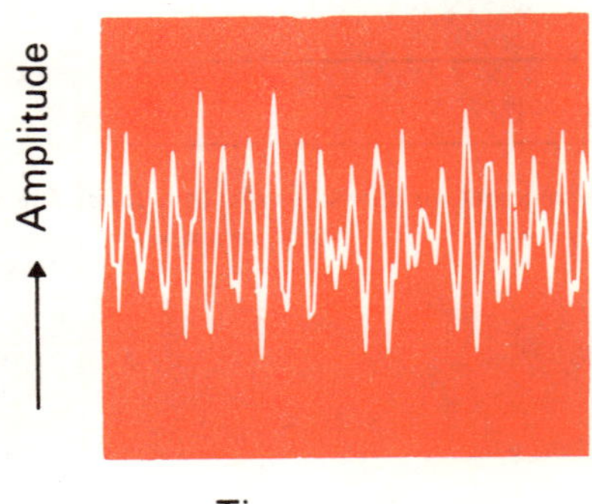

Figure 8b An aperiodic wave (a typical waveshape of the speech sound 'sh').

The spectrum

So far, we have considered only sinusoidal sound waves. A piston driven by a rotating wheel moves sinusoidally, so do the prongs of a tuning fork. For these the waveshape of the corresponding sound wave is also sinusoidal. Sound waves generated by our vocal organs, however, are almost never sinusoidal. In later sections, we will see several examples of speech waveshapes; in this section, we give only two examples. Fig. 8a shows the typical waveshape of the sound 'ah' and Fig. 8b shows the waveshape of the sound 'sh'. Although the waveshape in Fig. 8a is complicated, it clearly consists of repetitions of the same basic shape. In Fig. 8b, on the other hand, there are no such repetitions. The repetitive wave of Fig. 8a is called a *periodic* wave; Fig. 8b shows an *aperiodic* wave. Strictly speaking,

periodic wave
aperiodic wave

only waves with an infinite number of repetitions are periodic. But, in practice, many speech sound waves have enough repetitions to be regarded as periodic.

The waveshapes in Figs. 8a and 8b are extremely complicated and seem difficult to describe. Fortunately, Joseph Fourier, a French mathematician of the nineteenth century, showed that any non-sinusoidal wave, no matter how complicated, can be represented as the sum of a number of sinusoidal waves of different frequencies, amplitudes and phases. (The phases of the sinusoidal waves refer to their relative timing—whether they reach the peaks of their vibrations at the same time, for example.) Each of these simple sinusoidal waves is called a component.

Fourier analysis

phase

Fourier's results have been of great importance in analysing many physical phenomena, not only sound; they were, in fact, originally derived in connection with problems about heat flow in material bodies.

The *spectrum* of the speech wave specifies the amplitudes, frequencies and phases of the wave's sinusoidal components.

spectrum

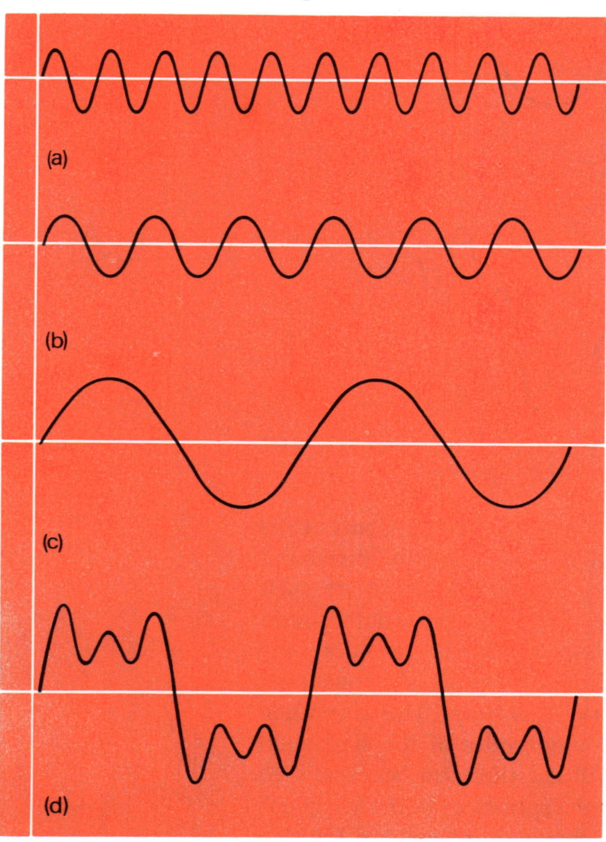

Figure 9 *Building up a complex wave: (a), (b) and (c) are sinusoidal components at different frequencies—(a) has five times and (b) three times the frequency of (c); (d) is the non-sinusoidal sum of (a), (b) and (c).*

The illustrations in Fig. 9 show the sum of three sinusoidal waves which together are the equivalent of a non-sinusoidal wave. The frequencies of the sinusoidal waves in Figs. 9a and 9b are, respectively, five and three times the frequency of the wave in Fig. 9c. Their amplitudes are half that of Fig. 9c, and they are 'in phase' at time $t = 0$. (All have zero value at $t = 0$.) When these three waves are added together—just by adding the displacements of all three, instant after instant—we get the clearly non-sinusoidal wave of Fig. 9d. Notice that the basic pattern of the non-sinusoidal wave repeats with the same periodicity as the lowest frequency component which is called the *fundamental*. The higher frequency components, when they are multiples of the fundamental, are called the *harmonics*.

fundamental frequency

harmonics

Figs. 10a to 10c show the same sinusoidal components as Figs. 9a to 9c, but the phase of the fundamental component in Fig. 10c is different from the phase of the fundamental component in Fig. 9c. The sum of the three components is shown in Fig. 10d. We notice that the phase-change of the fundamental alters the waveshape of the resulting wave. This shows that

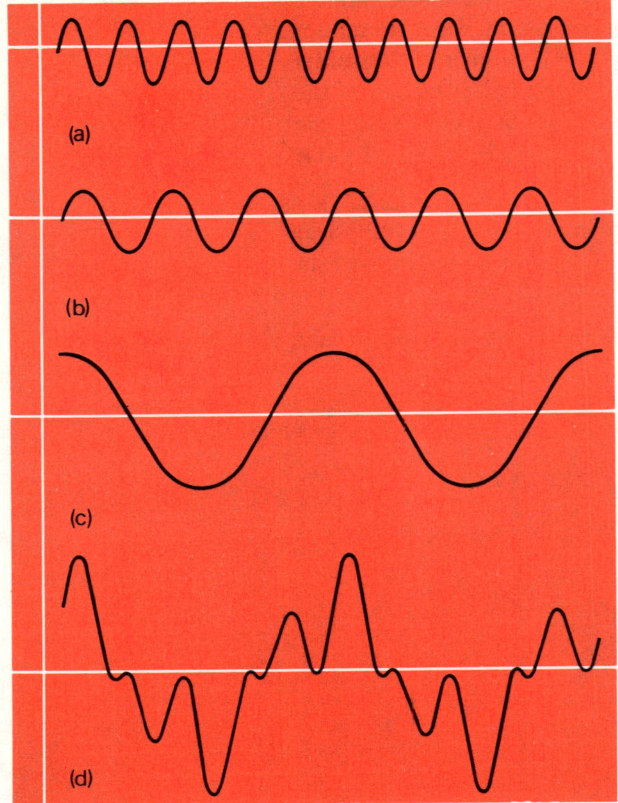

Figure 10   *The same component waves as shown in Fig. 9, but with the phase of Fig. 9 (c) changed. The resulting complex waveform has changed as shown in (d).*

we can get a variety of waveshapes by adding sinusoidal components of the same amplitudes and frequencies but of different phases. It is interesting that our hearing mechanism cannot always detect the effect of such changes. Indeed the wave patterns of Figs. 9d and 10d would sound indistinguishable. For this and other reasons, we usually consider only the 'amplitude' spectrum (that is the amplitudes of a range of frequencies) of non-sinusoidal waves and not the 'phase' spectrum. The amplitude spectrum specifies just the frequencies and amplitudes of the sinusoidal components. In the rest of this unit, we will use the term 'spectrum' to refer to the amplitude spectrum alone. Thus the spectra of both the signals in Figs. 9c and 10c are the same and are as shown graphically in Fig. 11: the *fundamental* at a frequency of 1 (arbitrary units) and the third and fifth harmonics with half its amplitude.

The graph of the spectrum of a sine wave is thus a single line of height proportional to its amplitude. The graph of more complex but periodic

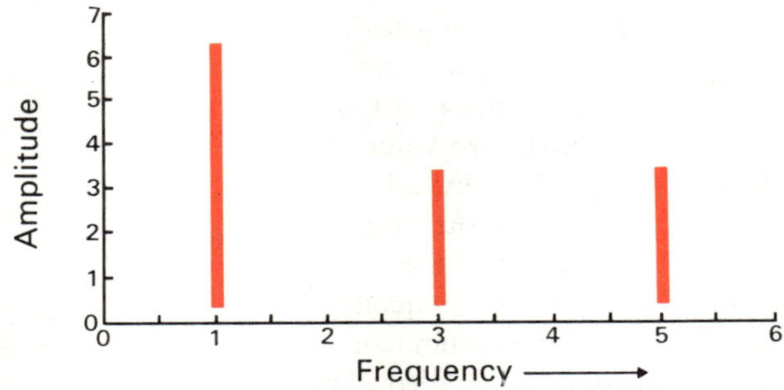

Figure 11   *The spectrum of the complex waves shown in Figs. 9 (d) and 10 (d).*

waves can, following Fourier's work, always be analysed into the sum of various sinusoidal waves, *all harmonics of the fundamental.* Thus *periodic signals always have a 'line spectrum'* (i.e. a spectrum consisting of lines at harmonic frequencies, as in Fig. 11).

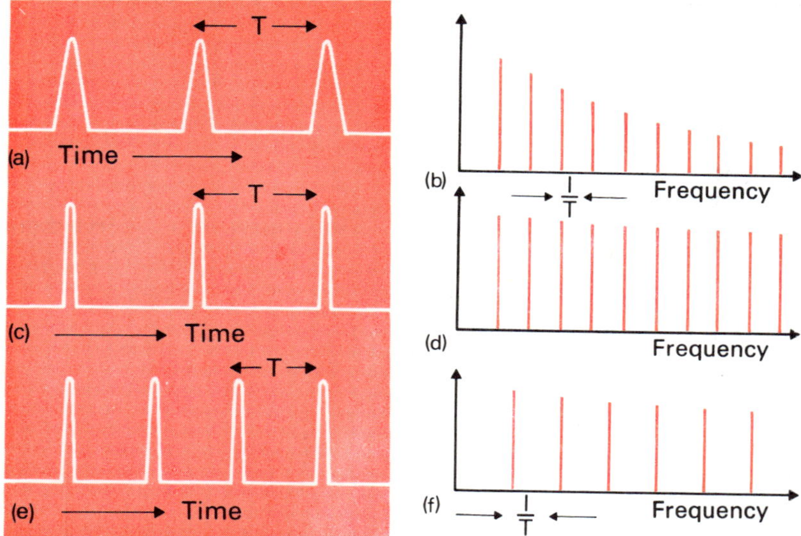

Figure 12   Periodic pulse sequences and their corresponding line spectra. (a) and (b) Broad pulses giving a spectrum whose amplitude decreases at the high frequencies. (c) and (d) Narrow pulses showing almost constant spectrum amplitudes. (e) and (f) Higher frequency pulses producing fewer spectral lines.

Of particular importance to the understanding of speech is the spectrum produced by the vocal cords. When we sing a steady note our vocal cords open and close at a constant frequency and let through a string of pulses, or puffs, of air, as indicated in Fig. 12a. Such a sequence of air pressure pulses has a line spectrum like that of Fig. 12b. It contains many harmonics, but their amplitude decreases as the harmonic number, and therefore the frequency, increases.

A sharper string of pulses, like that of Fig. 12c, produces even more harmonics and of more constant amplitude, as indicated in Fig. 12d. Differences in the pulse shapes produced by different voices account to some extent for differences in the voice quality of different speakers—even when they are singing the same note. Of course, when we speak our voice frequency alters with *intonation.* A higher frequency tone, like that of Fig. 12e will have a line spectrum of more widely-spaced harmonic frequencies.

A single tone, like 'middle C', or even a sequence of tones, like a tune, is obviously not speech. *Speech is produced by modifying the amplitudes of the harmonics produced by the vocal cords as they pass through the throat, mouth and nose.* These modifications occur because the throat, mouth and nose are resonant cavities and therefore, as discussed earlier, they respond better to some frequencies than to others. But let's study this in a little more detail.

2.3   Acoustic resonance and speech production

Enclosed volumes of air can resonate just like the pendulum we described earlier. When a sound wave reaches a volume of air enclosed in a container (e.g. the mouth or nose) an increase in the sound pressure compresses the air in the container. The 'springiness' of the air inside the container tends to push the compressed air out again. If the rarefaction of the sound wave reaches the container at the same time as the compressed air is being pushed out, the pressure of the sound wave and the pressure of the compressed air will add together and the air particles will move with increased amplitude. If the rate of arrival of the sound wave's compressions and rarefactions (the rate being equal to the sound wave's frequency of vibration) corresponds to a natural frequency of the enclosed air, we get increased movement or resonance. When we fill a bottle with water, we can actually *hear* it filling up. Resonance explains this: the splashing water generates sounds of many different frequencies, but the resonance of the air column above the water level emphasizes only those frequencies in the sound that are near its own natural frequency. As the bottle fills up, the size of the air column decreases (this increases the column's resonant frequency), and higher frequency components of the 'splashing' are emphasized. We know from experience that when the pitch of the sound from the bottle is high enough, little air is left in the bottle and it is time to turn off the tap.

The simple pendulum combination has only one resonant frequency: columns of air have many different resonant frequencies. We will consider only the resonances of tubes whose cross-sectional dimensions are small compared with the wavelengths of the sounds applied to them. The vocal tract is just this sort of tube for the frequencies primarily used in speech.

A tube with uniform cross-sectional area throughout its length has regularly-spaced resonant frequencies. The values of these resonant frequencies depend on the length of the tube. Sound waves developed in such a tube follow the rule that the wave must have zero amplitude at the closed end, and maximum amplitude at the open end. The lowest resonant frequency of this tube is the frequency of a sound wave whose wavelength is four times the length of the tube. The value of the tube's other resonant frequencies will be odd numbered multiples (three times, five times, etc.) of this lowest resonant frequency as shown in the margin diagram. When the cross-sectional area varies along the tube's length, the resonant frequencies are no longer uniformly spaced. They are spaced irregularly, some close together and some far apart, depending on the exact shape of the tube.

The human vocal tract is about 17 cm long and—at least when it produces vowel sounds—we can regard it as closed at one end and open at the other (the lips). The lowest resonant frequency of a uniform tube this long is 500 Hz; its other resonant frequencies are 1 500 Hz, 2 500 Hz, 3 500 Hz and so on.

Now when the string of air pressure pulses is produced by the vocal cords, it is transmitted outwards towards the lips. The vocal tract responds better to those components of the vocal cord puffs that are at or near its natural frequency. These components will be emphasized and the spectrum of the sound emerging from the lips will 'peak' at the natural frequency of the vocal tract. This is the process illustrated in Fig. 13. Fig. 13b shows the spectrum of the vocal cord output and Fig. 13c shows the frequency response of a simple resonator such as an organ pipe. Figs. 13d and 13e are the waveshape and spectrum of the sound wave produced when the sound in Fig. 13a is transmitted through the resonator of Fig. 13c.

Fig. 13c shows only one natural frequency, but the vocal tract has many, though there are three main ones, due to the mouth, throat and nose. The vocal resonators, therefore, emphasize the harmonics of the vocal cord wave at three principal frequencies, and the spectrum of the speech wave will have a peak for each of these natural vocal-tract frequencies. The values of the natural frequencies of the vocal tract are determined by its shape. Consequently the amplitudes of the spectral components will peak at different frequencies as we change the shape of the tract. Fig. 14 shows the spectra of sounds produced for three different vocal tract shapes.

Resonances of the vocal tract are called *formants*, and their frequencies, the *formant frequencies*. Every configuration of the vocal tract has its own set of characteristic formant frequencies.

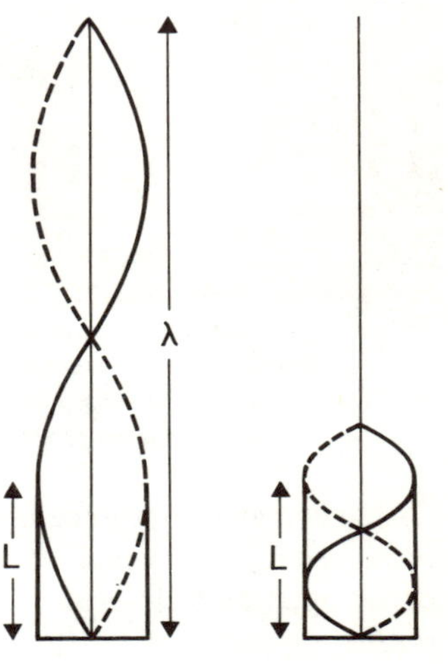

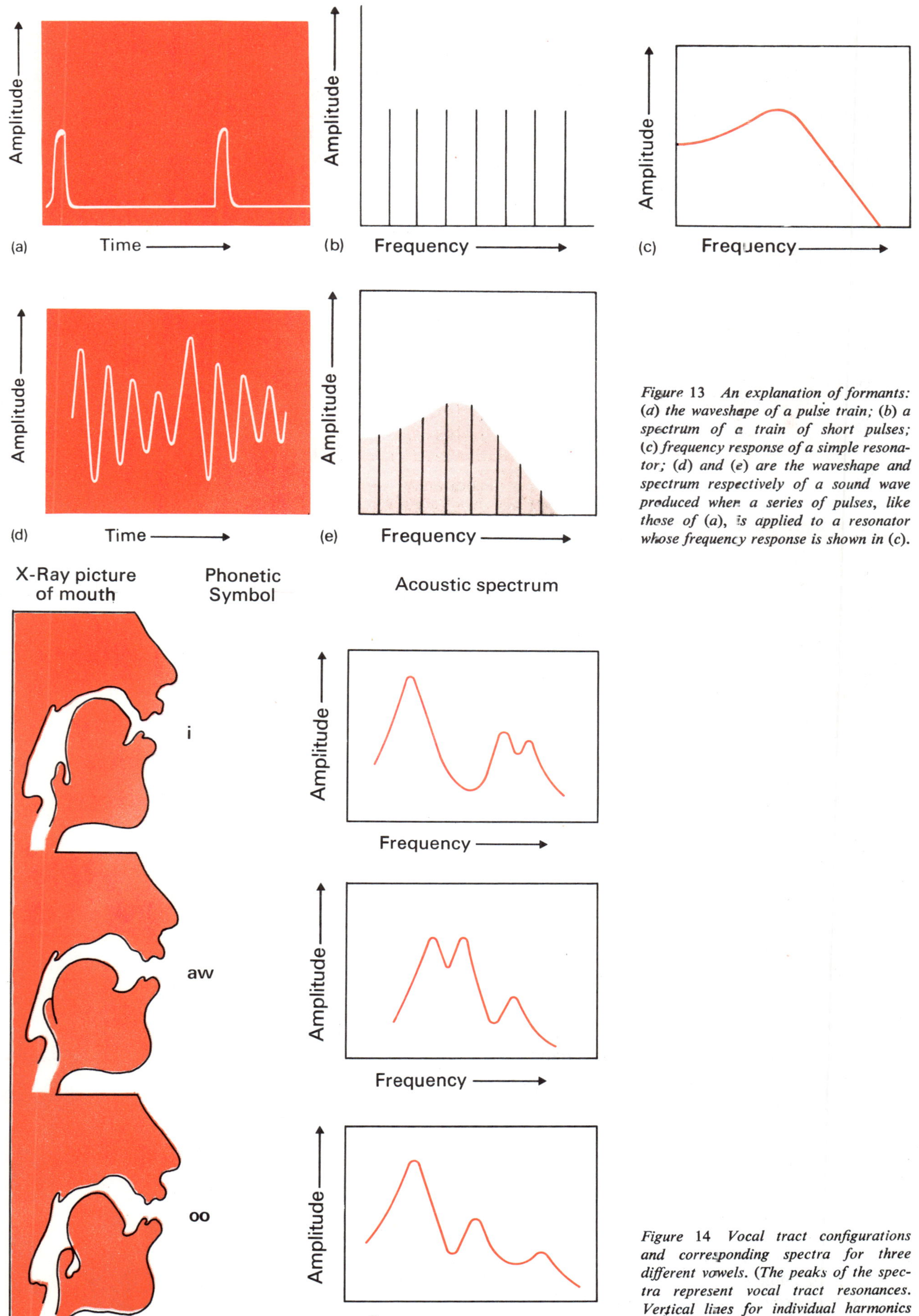

Figure 13 An explanation of formants: (a) the waveshape of a pulse train; (b) a spectrum of a train of short pulses; (c) frequency response of a simple resonator; (d) and (e) are the waveshape and spectrum respectively of a sound wave produced when a series of pulses, like those of (a), is applied to a resonator whose frequency response is shown in (c).

Figure 14 Vocal tract configurations and corresponding spectra for three different vowels. (The peaks of the spectra represent vocal tract resonances. Vertical lines for individual harmonics are not shown.)

In general, the frequencies of the formants will not be the same as those of the harmonics, although they may coincide. After all, there is no reason why they should agree. The formant frequencies are determined by the vocal tract, the harmonic frequencies by the vocal cords, and the vocal tract and vocal cords can move independently of each other. The independence of vocal cord and formant frequencies is illustrated in Fig. 15.

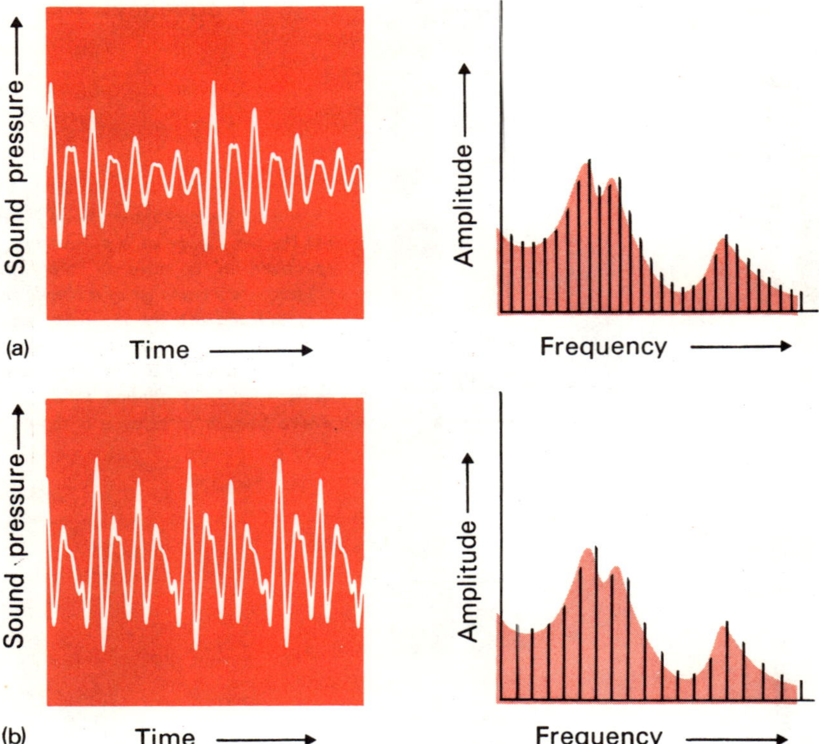

Figure 15   *The waveshapes and corresponding spectra of the vowel 'ah' pronounced with two different vocal cord frequencies: (a) vocal cord frequency equals 90 Hz; (b) vocal cord frequency equals 150 Hz.*

Fig. 15a shows the waveshape and spectrum of the sound 'ah' produced with the vocal cords vibrating at 90 Hz, and Fig. 15b the waveshape and spectrum of the same sound with the cords vibrating at 150 Hz. Even though the frequencies of all the harmonics have changed, the frequencies of the formants (and of the spectral peaks) are unaltered because the shape of the vocal tract remained the same.

It is important to realize, as illustrated in Fig. 13, that resonators can be excited by harmonics of a signal, if they are at the right frequency, not just by the fundamental frequency. Thus the mouth, which may have a formant resonant frequency of 1 500 Hz, can be made to resonate by the 15th harmonic of a 100 Hz signal (if it has a pulse-like waveform) or by the 5th harmonic of a 300 Hz signal. Sine waves have no harmonics so a sinusoidal signal of 300 Hz will only excite a resonator of 300 Hz.

SAQ 1

Consider again the simple pendulum with movable support. You recall that a large amplitude can be stimulated by small sinusoidal oscillations of its support at the natural resonant frequency of the pendulum.

(a) If these oscillations were impulses, across and back, as depicted in Fig. 12a, instead of sinusoidal oscillations, but still at the resonant frequency of the pendulum, would the pendulum's amplitude build up as before or not?

(b) If these impulses were at one-half or one-third the pendulum frequency would oscillation build up?

(c) What if the impulses were twice or three times the pendulum's natural frequency (i.e. two or three impulses per swing of the pendulum)?

(d) If the pendulum support were moved sinusoidally but at a half or a third the natural frequency would the oscillation build up?

For answer, see end of unit.

It is worth remembering at this point that we have now discussed resonance in two ways—first, as the natural frequency of a vibrator, usually damped because of energy loss in friction, etc. (as in Fig. 2b), and second, as a response to various input frequencies, which emphasize just one frequency. These two descriptions are, of course, complementary descriptions of the same phenomenon. Fig. 15a illustrates the equivalence well. The time graph looks like a repeated sequence of damped oscillations—a bit distorted because of multiple resonances—and the frequency graph shows the resonant frequencies.

SAQ 2

Figs. 2b and 2c on the one hand and the two curves of Fig. 3, on the other hand, are alternative descriptions of two somewhat different resonant systems. Which curves of each figure correspond to each other and why?

For answer, see end of unit.

Telecommunications signals

This description of complicated signals, in terms of their spectra, is also helpful in describing the signals needed for telecommunication such as in telephones or radio.

We could describe speech as the combination of two effects. We produce a string of pulses of air pressure by vibrating the vocal cords and we modulate the amplitudes of these pulses and their harmonics by making the pulses pass through resonators. The pulses themselves are unimportant as far as the words uttered are concerned, but without them (or something similar) the resonances, which form the vowels, would not be easily recognizable. So the pulses can be called the *carrier* and the resonances can be called *modulation of the carrier*. When we whisper we use a different *carrier*—a hissing signal. When children speak they use a higher frequency carrier. On each, modulation patterns can be superimposed to convey information so that the carrier is not just a meaningless tone or hiss.

carrier signal
modulation

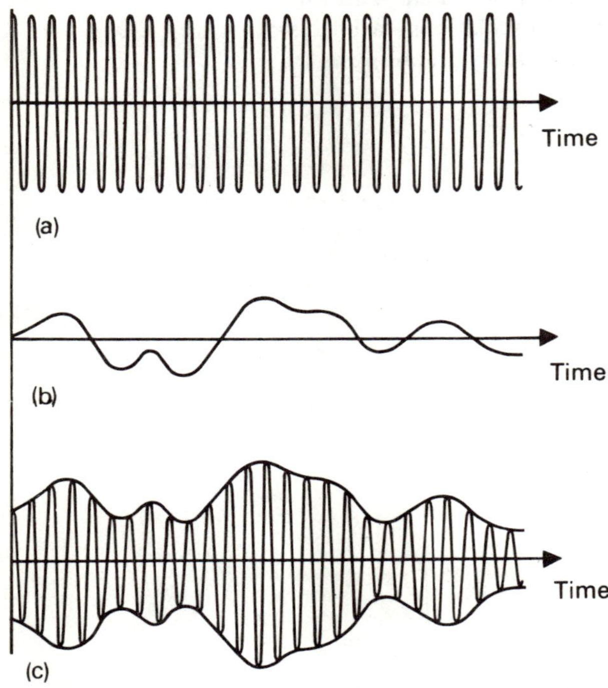

Figure 16   Amplitude modulation. (a) A high frequency sinusoidal carrier; (b) a message signal: for example, part of a speech waveform; (c) the carrier amplitude modulated by the message signal.

The same idea is used in radio, but the details are quite different. A single sine wave is used as the carrier in this case, and the information, the music or the speech, is superimposed on this carrier by varying its amplitude (Fig. 16). This is called amplitude modulation. So tuning in to a radio station involves adjusting the frequency of a resonator (this time made from electronic components rather than a pendulum or a volume of air) so that it is excited by the carrier signal. This happens when the two have the same frequency. Thus the radio is 'tuned in' to the radio transmission. The important point, however, is that once the radio or TV receiver is tuned in to the carrier, all the other transmissions at different carrier frequencies are reduced to relatively negligible amplitude, and the message carried by that one resonant signal can be extracted from it by following its variation in amplitude. This is called *demodulation*.

amplitude modulation

demodulation

So by attaching each message to a different carrier it is possible to transmit many messages simultaneously through the air (by radio) or along a wire (in telephones) and separate the output at the receiver. This is called multiplexing. All you need in principle is a resonator for each carrier frequency, followed by a demodulator. The carrier frequencies used for television transmission by satellite (see the case study T100/BK2) are in the range of 800 million to 10 000 million Hz. Sound radio uses carriers of around 1 million Hz. Telephone transmission along cables uses carriers of from 12 thousand to 1 million Hz or more. The carrier can be almost any frequency above the frequencies used in the message—the principle is always similar.*

multiplexing

Another important fact about speech, music and television which analysis, in terms of frequency, reveals is the *range of frequencies* necessary for transmission of these signals. This range is called the *bandwidth*.

bandwidth

If we look back at Fig. 15b the spectrum there shows the harmonics of a fundamental frequency of 150 Hz. It shows seventeen harmonics so the highest frequency shown is $17 \times 150 = 2\,550$ Hz. If we consider the frequency content of all speech sounds, including consonants such as s, th, f, etc., we find we need a range of frequencies from about 100 Hz to about 3 400 Hz for good reception. (This will be demonstrated in a radio programme.) If the higher frequencies are absent each s and f is lost and everything sounds woolly. If the lower frequencies are absent the speech loses its quality and individuality. Usually, therefore, telephones are designed to handle a bandwidth of 3 400 Hz for each conversation.

If the carrier is amplitude-modulated by a speech waveform as indicated in Fig. 16 then a bandwidth of 3 400 Hz on *each* side of the carrier is normally needed. If the carrier frequency is 1 million Hz, for example, then the required frequency range is 996 600 Hz to 1 003 400 Hz. This is called 'double side-band amplitude modulation'. If the message signal is only present on one side of the carrier, it is called single side-band modulation.

If each speech channel on a trunk telephone line using single side-band modulation occupies a bandwidth of 4 kHz, how many simultaneous conversations can be transmitted in the normal 'trunk' line covering the frequencies 60 to 108 kHz?

12 channels.

Television transmission requires a much larger bandwidth than speech. We can quite easily work out how much, as follows: Think of a typical 625 line television display. The number of lines is chosen so that the lines are just invisible from a normal comfortable viewing distance (as discussed in the previous unit). By the same reasoning it follows that there should be about 800 points to each horizontal line of the television scan because the picture is wider than it is high. So each picture can be thought of as made up of $625 \times 800 = 0{\cdot}5$ million separate points, and the appropriate brightness of all these points is the information contained in the TV transmission. As there are 25 different pictures transmitted every second it follows that the rate of transmission must be 12·5 million picture points per second. Now the most detailed picture which it is possible to send is one in which each picture point is of the opposite brightness to the one which preceded it. In other words the picture is just a fine mosaic of black, white, black, white, etc. So to transmit this most demanding picture we need a signal which *alternates* at a frequency of 6·25 million cycles per second (each cycle being black followed by white). A sinewave of this frequency is the simplest suitable signal.

TV bandwidth

* Different methods of modulation can be and are used but most methods in use at present involve the idea of a sinusoidal carrier. Other methods more closely resemble speech in that the carrier signal is a periodic pulse pattern.

At the other extreme, a picture which is all white or all black requires a signal which varies very slowly, in fact 25 cycles per second, so the bandwidth needed for television is about 25 Hz to 6·25 MHz,* nearly 2 000 times the bandwidth needed for speech.

High quality radio transmissions and gramophone records need a wider bandwidth than speech (though speech too is also improved with a wider bandwidth) but still much less than television. A bandwidth of 15 kHz is usually provided and is adequate for all but the most discriminating.

One of the telecommunication possibilities of the future is a telephone with a small television display alongside (the Videophone or Viewphone). If this television display has a 300 line horizontal scan, 25 times a second and if the picture has the proportions, height : width = 3 : 2, what bandwidth will be needed for transmission?

About 750 kHz.

The importance to technology of this discussion about bandwidth is primarily this, that every telephone conversation and every television signal must be provided with its appropriate bandwidth, and if they are all to be transmitted simultaneously, whether by radio or by a wire cable, they must each have a different carrier frequency well enough separated in frequency from its neighbour, otherwise they will interfere with each other—unless physical locations of the transmitters (or cables) are sufficiently far apart. It is because the number of telephones and telephone calls is continually increasing that the demand for telephone carrier frequencies is always rising. At present, there are about 130 million telephones in the world and there will be 10 million more next year. Every ten years more telephone equipment is manufactured than ever existed before. It is for this reason that satellites like Intelsat IV are used (see T100/BK2). They relieve the pressure of demand on the ground for enough 'channels' of radio communication because some transmissions can be sent that way instead of over the surface of the ground. Other ways for the future might involve the use of lasers to convey signals accurately from one place to another without interference with each other, possibly along glass fibres.

satellites

* The abbreviation MHz (for megahertz) means 'millions of cycles per second'.

Section 4

Some further properties of speech and its production

In the last section we discussed a number of important ideas associated with signal properties analysed in terms of frequency. I now want to use these ideas to help describe speech in a little more detail so that the subtlety and complexity of speech and its interpretation can be more readily appreciated. The main point I want to convey is that speech is an extremely varied pattern of sound. The same words uttered by two speakers are usually carried by very different signals. Secondly, despite this and despite great variation in the characteristics of the listener and his environment—which may distort or add noise to the signal—speech patterns can be readily understood. A communication system so resistant to variation of signal, transmitter and receiver is at present unknown to technology and may have much to teach us about the design of future systems.

4.1 Articulation

Adjusting the vocal tract's shape to produce different speech sounds is called *articulation*. The individual movements of the tongue, lips and other parts of the vocal tract are called articulatory movements.

articulation

We see, then, that air flow from the lungs provides the energy for speech wave production, that the vocal cords convert this energy into an audible buzz and that the tongue, lips, palate, etc.—by altering the shape of the vocal tract—transform the buzz into distinguishable speech sounds.

Because people's heads are different sizes, and their mouths, noses, throats, lips, tongues are different shapes, the sounds uttered when they say the same words are quite distinct. Indeed voice spectra are being developed to enable the police to recognize people by their voices.

Two other methods are available for making the air stream from the lungs audible. In one, the vocal tract is constricted at some point along its length. The air stream passing through the constriction becomes turbulent, just like steam escaping through the narrow nozzle of a boiling kettle. This turbulent air stream sounds like a hiss and is, in fact, the hissing or *fricative* noise we make when pronouncing sounds like 's' or 'sh'.

The other method is to stop the flow of air altogether—but only momentarily—by blocking the vocal tract with the tongue or the lips, and then suddenly releasing the air pressure built up behind this block. We use the 'blocking' technique to make sounds like 'p' or 'g' which are called *plosives*.

4.2 Linguistic organization

The variability of speaker characteristics does not end with the sound patterns. The message to be transmitted from speaker to listener is first arranged in linguistic form; the speaker chooses the right words and sentences to express what he wants to say. The information then goes through a series of transformations into physiological and acoustic forms, and is finally reconverted into linguistic form at the listener's end. The listener fits his auditory sensation into a sequence of words and sentences, and the process is completed when he understands what the speaker said.

The most familiar language units are words. Words can be thought of as sequences of smaller units, the speech sounds or *phonemes*. Vowel sounds and consonant sounds are phonemes. There are about 44 of them in the English language—though fewer in American English! Words may contain any number of phonemes, usually between one and ten. It is quite surprising to realize that at a normal rate of speaking—say 200 words per minute—phonemes are uttered at about 1 000 per minute or between 15 and 20 per second. This is about the speed at which visual flicker is easily noticeable—yet in speech the 44 different phonemes can apparently be recognized at this rate.

Now words are, of course, combined into longer linguistic units, *sentences*. The rules that outline the way sequences of words can be combined to form acceptable sentences are the *grammar* of a language. Grammar tells us that the string of words 'the plants are green' is acceptable, but the sequence 'plants green are the' is not.

Grammar alone, however, does not determine word order. Sentences must *make sense* as well as satisfy the rules of grammar. For example, a sentence like 'the horse jumped over the fence' is both grammatically acceptable and sensible. But the sequence 'the strength jumped over the fence', although grammatically correct, is meaningless and does not occur in normal use. The study of word meanings is called *semantics*, and we can see from our two examples that word order is influenced both by grammatical and semantic considerations.

Stress and *intonation* are also part of linguistic organization. They are used to express such things as the speaker's emotional attitude, to make distinctions between questions, statements and doubt, etc., and to indicate the relative importance attached to different words in a sentence.

Now all this richness in structure of a language still allows a tremendous freedom in language organization. Although there are constraints embodied in the semantics, syntax (grammar) and the way phonemes and sounds are strung together, sentences uttered by different speakers to express the same idea are rarely the same at any level. The words used are different, the sounds constituting the words are different. One speaker is slow and hesitant, inserts 'ums' and 'ers', another gabbles. One is verbose, another is very abrupt. Somehow, however, the idea gets across from the speaker to the listener, not just in a quiet sitting room but in a noisy car, train or canteen.

Although at the present time, until we know how to analyse speech better we have no way of expressing *how much* variation there is between the linguistic styles of different speakers, we can nevertheless express, to some extent, the differences in the sounds they produce.

A considerable amount of spectral information is available about individual speech sounds, particularly for the vowels.

The most significant features of the vowel spectrum are the frequencies and amplitudes of the various formants. These, you will recall, correspond to the resonances of the vocal tract, and they produce peaks in the speech spectrum. Even though a certain amount of ingenuity is sometimes required to identify the spectral peaks caused by vocal tract resonances, the spectrum is our best guide for finding these important formants.

Usually, the first three formant frequencies are adequate for recognition. These are given in Fig. 17a for ten English pure vowels. The values shown are the average frequencies obtained from the speech of a number of male and female speakers. The vowels were pronounced in one syllable words, like 'heed' and 'hid' and each syllable was spoken in isolation. The formant frequencies for women are higher than for men because, on average, women are smaller than men. It has nothing to do with the pitch of their voices. The formants for children are even higher, again because their faces are smaller.

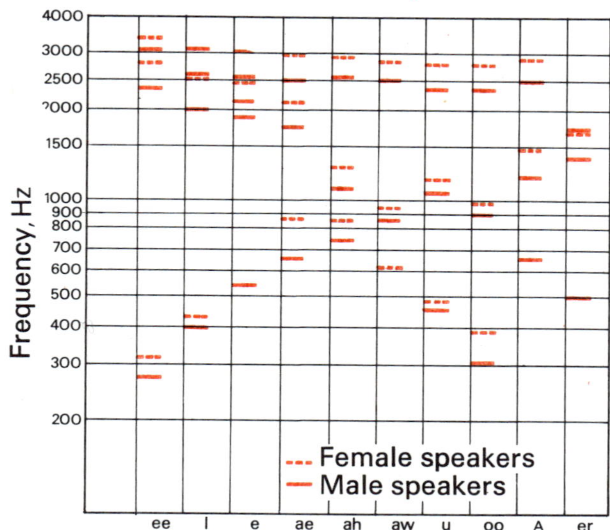

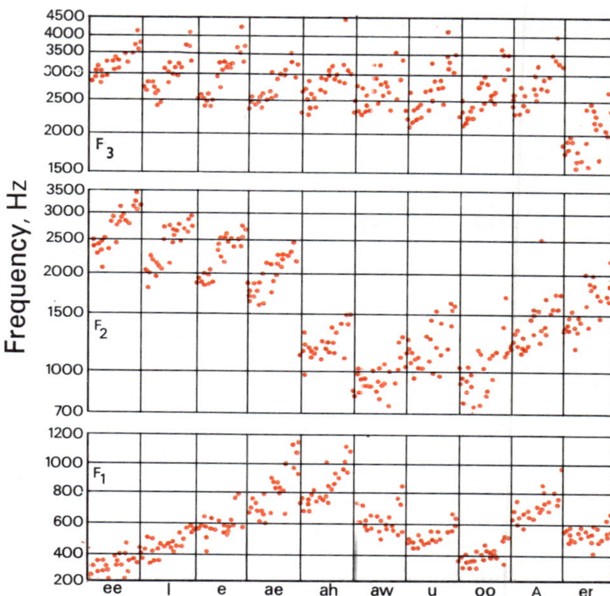

Figure 17a   *Average values of formant frequencies for ten English pure vowels. The vowels are spoken in isolated single syllable words.*

Figure 17b   *The formant frequencies of ten English pure vowels as pronounced by a number of different speakers. (The notations, F_1, F_2 and F_3 refer to the first three formant frequencies.)*

The variability of the formant frequencies is shown in Fig. 17b, where individual results are plotted, rather than averages. We see that the range of formant frequencies produced when any one vowel is uttered overlaps the range of adjacent vowels. Closer investigation shows that this overlap maintains itself even when we consider combination of first and second formants.

So how, in fact, can we distinguish adjacent vowels? No one really knows yet. Actually, the variation is much greater than that shown in Fig. 17b because speech is continuously varying. There is no moment in normal continuous speech when the formant frequencies hold steady, waiting to be recognized so to speak, as implied in Fig. 17b. The pattern changes rapidly and rushes on. Indeed the changes are as significant, if not more significant than the frequencies themselves for they enable us to predict what the next sound is likely to be. They also help us with recognizing consonants. To form the consonant 'g', for example, it is necessary to press the tongue to the roof of the mouth. But this movement also alters the shape of the resonant cavities formed by the mouth and throat. So by observing the *variation* of the formants it is possible to anticipate the consonant which is about to emerge. Somehow, as listeners we humans can do this effortlessly, almost, it seems, instinctively, and with a skill quite unconnected with what we normally call intelligence.

It is this remarkable facility we have which perhaps holds lessons for us in engineering design. So let us now see how we can begin to study this problem.

Design strategy for engineering systems

We have considered many of the known physical characteristics of speech and we might therefore imagine that it is quite a small step from this knowledge to, for example, constructing a machine to recognize speech. With such a machine it would be possible to type letters or articles directly from dictation without the intermediary of a typist, or to instruct computers and other automatic machinery without first learning some other code such as typing, or a computer language such as FORTRAN or BASIC.

Unfortunately, despite many attempts, machines with only the most rudimentary pattern recognition ability have been built. Typically they are capable of recognizing fairly successfully only a few words spoken by only a very few speakers. The trouble is that speakers differ so much in speed of speaking, intonation, accent, emphasis, physical characteristics, frequencies used and so on, that machines keep making mistakes. Yet, somehow, human beings can see past all this distortion and variation to the underlying words and recognize what is being said with remarkable accuracy.

This human achievement is one which neither science nor technology has yet studied in any great detail (partly because it is so difficult to experiment with). Perhaps the key is to understand how to use 'redundancy', either in equipment or in the signals handled, to achieve the freedom from error we need. The rest of this unit is devoted to a study of what this concept of redundancy means and how it is being exploited, albeit in very simple ways, at the present time.

5.1 Redundancy in engineering systems

The word redundancy usually means something like 'not wanted', at least where employment is concerned. In this section we are going to use the word to mean 'not always needed' rather than 'not wanted'. In other words, an engineering system which contains redundancy is one that has spare parts built into it which can automatically come into operation, or remain in operation, when some other parts fail, so that the system is not significantly impaired. For example, most pylons or steel truss bridges do not collapse if any one of the members fails, indeed the bridge's performance will normally be unaltered, but it will not be able to stand up so well to extremes of overload or strain. What happens is that the remaining trusses carry a different distribution of load. Thus the structure is redundant in that it contains more trusses than are strictly needed. Thus redundancy is a kind of insurance against failure. This is the sense in which the term is being used here.

In most dynamic pieces of engineering such as machines, as distinct from static structures like bridges, a different strategy is adopted to deal with failure. Any faulty parts are either repaired or replaced or sent away for servicing. Either way the machine goes out of action for a period. Furthermore a computer, radio set or washing machine usually becomes useless if any *one* of its parts fails. No means is provided to ensure that the machine carries on working, even if inadequately, after a part has failed, and

redundancy

failure

certainly no method is provided for automatically switching in a new part when one of the original parts fails.

It could be argued that the engines of motor cars contain a certain measure of redundancy, in that a four-cylinder engine will still continue to propel the car even when one cylinder has ceased to function properly, but here even the *normal* performance is impaired by failure, not just the performance under extreme conditions, so it is not true redundancy. The two independent braking systems being provided in more and more modern cars are more nearly what we are looking for. If one system fails the other automatically takes over. But even here normal behaviour is not unaltered in the event of partial failure.

Where aircraft or manned satellites are involved, the degree of redundancy is greatly increased, since the consequences of a single failure in such complex pieces of machinery cannot be allowed to lead to loss of life. Similarly aircraft with three or four engines are preferred to aircraft with one or two engines, since such a design leaves greater room for engine failure without disastrous consequences. Indeed an aircraft is required to be capable of meeting all important performance specifications with one engine out of action. However, in general it is true that redundancy in engineering systems is very rudimentary.

In living organisms exactly the opposite strategy is the general one. It is normally (except with humans) impossible for a living organism to arrange to 'go to the doctor' and have itself repaired. Instead it has evolved to tolerate a large amount of failure during its lifetime without significant overall system impairment.

There appear to be two main ways in which this has been achieved. First, by making the organism self-repairing. Second, as in the central nervous systems where replacement of dead cells is impossible, the system is structured so that a large number of parts can fail before the system ceases to function. The human brain, for example, contains about 10 000 million cells, of which about 10 000 die every day. Many of these cells undoubtedly contain memory traces, and these memory traces are lost for ever once the cell has died, for there is no regeneration of cells in the central nervous system. Nevertheless the brain appears to continue to function satisfactorily at least for 60 to 70 years, so the lost memory traces must presumably be held elsewhere in the brain as well. In old age senility begins to set in, at which time perhaps between two per cent and ten per cent of the brain has died and been cleared away.

The same power of recovery seems to exist in the environment, which has achieved a measure of balance between synthesis, decay, regeneration and degeneration, which is overall fairly static, and in which species of organism can progressively evolve over millions of years. Only recently are we discovering that this power of recovery, this redundancy of self-repairing systems, is not wholly resistant to pollution and destructive influences.

5.2 Redundancy in speech

There is a similar kind of redundancy in the audible and visual signals which human beings habitually handle. We saw earlier how there is considerable variation between different speakers, both in the range of physical parameters that describe their different speech patterns even when they are saying the same things, and in linguistic structure generally.

It is useful therefore to think of a spoken sentence as a signal, which represents the message it carries, corrupted by noise or distortion. In order for this speech to be recognizable, despite all this distortion, the whole **distortion** signal must contain much that is redundant much of the time, so that if parts are lost because of this noise or distortion it is still possible for the message to shine through, so to speak, despite all the error occurring in the particular signal pattern.

Actually, of course, speech conveys more information than simply the sentences that are uttered. It also contains information about who the speaker is, about the mood the speaker is in, and even about what his true intentions are. Thus speech certainly contains a lot more information than is necessary simply to convey the words to the hearer. Nevertheless it is probably true that much of the information which is used to identify speakers or identify their mood can be used in some way to help correct any errors which might otherwise be made when a listener is trying to understand what the speaker is saying. In this sense it is redundant.

Now the point of this discussion about redundancy in living systems, as far as technology is concerned generally, is that the 'design strategy' used in the 'design' of living organisms seems at the present time to be quite different from the design strategy used in scientifically-based engineering systems. The purpose of engineering science is to analyse a problem into separate components so that a system or machine can be designed to meet a required specification with the minimum of cost. The basic strategy as far as living organisms are concerned seems to be to intertwine all the various functional parts into a complex structure so that if any parts fail the remainder can continue to handle the whole problem. The information handled in speech, or the functions achieved by the central nervous system, seem to be spread thinly over many parts of both the signal and the system so that when any part fails, there is always the rest to rely on. It seems to be true too, the deeper one studies the matter, that no serious loss of efficiency results from this strategy when many functions are involved. *Each acts as a source of redundancy for the error-free execution of the other functions.* It could be, therefore, that in the future the main strategy for the design of engineering systems should more nearly emulate that of living systems.

So let us see how some hesitant steps are being taken towards the design of systems and signals which make use of redundancy in the signal in order to achieve overall reliability in the technology of communication.

First we must establish beyond reasonable doubt that human communication contains a great deal of redundancy. We will then go on to try to define in greater detail what we mean by redundancy, and will finally discuss how redundancy can be added deliberately to the 'information' part of a signal in order to ensure that the information is received without error, even though the signal may itself contain error.

What we want to do first, then, is to look at evidence which demonstrates the presence of redundancy in human communication.

5.2.1   Bandwidth for transmitting words

For speech to be transmitted over the telephone system we need, as we saw earlier in this unit, a bandwidth of 3·4 kHz. In other words, the communication channel to handle this speech must be capable of responding to changes in the signal amplitude at a rate of at least 3 400 changes per second. Now a normal rate of speaking is about 160 words per minute, or a little under 3 words per second. If each word on the average contains five letters, then normal speech, translated into writing, involves a production of about 15 letters per second. We also know that it is possible to transmit letters of the alphabet using a simple two-level code, which uses five pulses per letter (see Table 3 in Section 7.1); thus 160 words per minute can be transmitted using pulses at a rate of 75 pulses per second, and this requires a bandwidth of less than 50 Hz. *So the words which are contained in speech can be transmitted using between 1/50th and 1/100th of the bandwidth that is used by speech itself.* Hence, as far as the words only are concerned, and ignoring what other information is involved, speech is nearly 99 per cent redundant. Only 1 or 2 per cent of the signal which constitutes speech is actually needed to convey the *words* it contains. Who the speaker was, his mannerisms, etc., are all lost.

5.2.2 Resistance to distortion

We can take a speech signal, such as that recorded on a tape, and mutilate it in a variety of ways, as is demonstrated in the radio programme related to this unit. Despite these very severe mutilations, which appear to remove almost all of importance from the signal, human beings are still able to understand it. From which it follows that what was removed from the signal must have been redundant in some sense or other.

5.2.3 Error correction using context

Earlier we discussed the structure of speech expressed in its grammar, semantics, phonetic sequences, etc. This structure can be used, of course, to correct detailed errors in a signal. We know, for example, that nouns follow definite or indefinite articles, and that verbs are usually associated with nouns, etc. We know too that ideas cannot be green, nor can pebbles be happy or furious, except figuratively. Thus the rules of morphology (word structure), syntax (sentence structure) and semantics (meanings) restrict the range of 'components' which go together to make acceptable English sentences. Rather surprisingly nearly all human beings, however intelligent or unintelligent they may be, seem to have a practical grasp of these (or similar) rules and are able to put them into effect when they utter or hear sentences. In this sense speech seems very basic to humans.

error-correction

Similarly most people who can read would have little doubt as to what was intended by the words 'it loohs as thouh it widl rain'. Rules of syntax and of semantics enable the reader to correct the spelling or typing errors which actually have occurred in the sentence. Such error-correction would not be possible if there were no redundancy.

So the signals humans use for communication, both written and spoken sentences, and the equipment they use to decipher them—mostly the central nervous system—are highly redundant. The result is that speech can be understood despite extreme distortion, appalling background noise and continuous and severe (in engineering terms) failure of parts of the apparatus used to deal with it. As engineering systems in the man-made world become more and more complex the need for this kind of system reliability increases.

5.3 System reliability by correcting errors and failures

To identify error and then achieve error-correction in a signal such as speech we must attempt to identify the information content of the signal and thus identify what is redundant. Once we have reduced the signal so that only essential information remains, we can then add redundancy to it according to clear rules. Then a receiver of the signal, which has been properly designed or instructed to use this redundancy, can correct errors which might occur when the signal is transmitted. So first we must identify the essential information in a signal and reduce it to its most elementary form. To do this properly with speech is as yet impossible, for it is much too complex a signal. However, some attempts can be made, as we shall see.

information

A simple way of identifying the information in a signal is to discover to what extent it is possible to guess the next part of the signal, given the signal so far transmitted. If any redundancy remains in the signal, a sophisticated receiver, such as a human being, will be able to guess, with a likelihood greater than that of pure chance, what the next part of the signal will be.

For example, with a random sequence of 1s and zeros like the first one shown below, the chance of guessing what the next digit will be is 50 per cent, or 1 in 2, because the sequence contains no pattern.

101100100011010111001011110100110110   Random sequence

100001000010000100001000010000   Periodic sequence

However in a sequence such as the second one, the chance of guessing the next digit is very high, which means that there is really no need to send the next digit, the receiver can construct it on the basis of what it has so far received and what it 'knows'. So the message in this signal is small and most of the signal is redundant.

The following experiment, which you can perform at home or together as a group, should make this point clearly:

Get a pencil and paper ready and then ask a colleague to open a novel at any page and select a paragraph of about ten lines. Then ask him to read you the first two lines of the paragraph. You should ask him then to allow you time to guess what the next letter is and write it down (a 'space' qualifies here as a letter). After you have made your guess, ask him to read you what the *letter* was and write that down above your guess. You then guess the next letter and again he reads you what the correct letter was. Proceed with this guess and statement alternately for as many lines of the paragraph as you can stand, recording your guesses and answers as you go. At the end of this exercise count up the number of correct guesses and divide it by the total number of guesses you made.

You will probably find that you guessed right more than half the number of times, even though you had 27 letters (including the space) to choose from. If there were no redundancy in the paragraph your chance of guessing correctly would have been only 1 in 27.

You can check this if you like with the following experiment:

Fig. 18 shows a random sequence of letters. Give this to your colleague and ask him to carry out the same procedure as before. Ask him to begin somewhere in the middle and go right round and back to that point so that you cannot inadvertently cheat by seeing what the first few letters are.

L M W space N L T W X A O U X K O W space
I H N J J R O Z J space K K Z E L G W V Y
N M H F G Z space D U V L L G H V K G V P
E F I B H Y A D M Q Q R R S R T P P T D T
C E H Y V A I C S F C T S C S S D Y M V Z
F Q O Z X E P X B N U

*Figure 18   A random sequence of letters in which each letter (including space) is equally
likely.*

In engineering terms this means that the information content of one letter of the novel, given the rest of the sentence, is much less than in a random sequence of letters (and spaces) chosen from the English alphabet. In engineering terms the aim is accurate communication of *anything*, not just meaningful sentences, so the fact that the random sequence *means* nothing is not important.

We have spent rather a lot of time demonstrating the existence of redundancy in engineering systems and in human communication, with the result that the concept of redundancy should by now be becoming clear. However, one complementary concept of information does require further explanation, since in technology the word 'information' does not mean quite the same as it means in everyday language.

Section 6

The concept of information in communication

The important difference between the concept of information in technological situations and its meaning in everyday language is that in technology the properties of the 'receiver' are involved in defining what is meant by information. (The receiver might be a human being, or some kind of electronic apparatus such as a computer which can store data.)

In everyday language it is sensible to talk about 'the amount of information' in a newspaper article or in a headline or in a telegram, because it is thought that the amount of information depends solely on what is written. In technology, however, it is the amount of 'instruction' conveyed to a receiver, whether it be a machine or a human being, which is important, so the amount of 'knowledge' the receiver already has bears upon the problem. Indeed, in many ways, it would be better if the term 'information' in technology were called 'instruction', since the information conveyed in the message is quite closely related to how much the receiver would not have been able to anticipate on the basis of his own prior knowledge.

In the first guessing experiment you carried out, each letter in the passage from the novel added little to what you already knew, so, in engineering terms, its information content was small. In the second passage, the random one, your knowledge of the structure of the language was useless (but assuming that it was English did help) so the information rate, per letter, was high—even though the passage was meaningless.

Thus the essence of the idea of information in engineering is that if the 'receiver' has a *good* chance of inferring the signal, then the information rate of the signal is low; whereas if he has a *poor* chance of inferring the signal then the information rate is high. Of course, a 'machine' cannot infer, but it can work out probability and extrapolate from previous data and rules provided it has a means of storing and utilizing records of previous events.

information rate

Let us take an example.

The number of letters in the English alphabet is 26. Thus there are $26 \times 26 \times 26 = 17\,576$ ways of forming 3-letter combinations of these letters. (Not words, just combinations like PZQ or PUL, etc.) The chance of guessing the *right* combination, given only that one is about to be sent to you is 1/17 576. *Therefore* the information content is related to the number 17 576. (Actually it is $\log_2$ 17 576 as we shall soon see.)*

Now in the English language there are far fewer 3-letter words than 17 576. Let us suppose there are 1 000 of them. If I am told an English 3-letter word is about to be sent to me, I now only have 1 000 to choose from, so I have a chance of 1/1 000 of guessing right. The information conveyed by receiving the message is now only related to 1 000 (i.e. $\log_2$ 1 000).

If I now start behaving like an intelligent human being, and have just heard part of a sentence, the chances are always quite high that I can infer the next word. If I am told it is a 3-letter one, I may be able to guess with very little equivocation (or uncertainty). Thus my chance of guessing right may be as high as 1/2 or 1/3; in which case the information content is related to 2 or 3 (actually $\log_2 2$ or $\log_2 3$). We are now back again to the guessing experiment, and if you guessed right half of the time, then you have shown that 'the information rate per letter' was about $\log_2 2$ or *1 bit of information*.

This measure, expressed as the 'logarithm-to-base-two' of the 'reciprocal-of-the-probability' of inferring correctly, is arrived at as described in the next section.

6.1   A measure of information

Suppose in a communication situation, an alphabet of eight letters is to be used, and we, the receiver, have no clues as to which of the eight is to be sent in any particular transmission. The probability of guessing is then 1/8. The information content is the smallest number of 'either–or' instructions we need in order to select a specific letter of the alphabet or, more generally, to select a specific item from a list of eight items. An 8-letter alphabet is set out in Table 1.

Let the first instruction tell us 'the required letter is in the top half of the list, if the digit sent is a 1, but otherwise not'. If the second instruction tells us 'the required letter is in the top half of the previously selected

Table 1

Alphabet	1st instruction	2nd instruction	3rd instruction
A		1	1
B			0
C	1	0	1
D			0
E		1	1
F			0
G	0	0	1
H			0

* If you are unfamiliar with logarithms, refer to Modelling I.

portion if the digit is a 1, otherwise not'; then, for example, the instruction 01 will, as indicated in the table, select E or F.

Three of these 'binary' instructions (that is, one offering only two possibilities) will similarly select from among eight letters. For example, 011 identifies E in the table.

What code selects the letter B? How many instructions are needed to select among 4 letters, 16 letters, 1 024 letters, 2^{20} letters?

The code for B is 110 and for the second question the answers are 2, 4, 10, 20. In general if the alphabet has n letters (so that the chance of guessing right is $1/n$), the number of binary instructions I is given by:

$$n = 2^I$$

or, taking logarithms to base 2, the 'information' is given by

$$I = \log_2 n \text{ bits per letter.}$$

The unit 'bit' is short for 'binary digit' of information. The formula thus says, in effect, that $\log_2 n$ binary digits are required to specify the identity of one item from a list of n items. The answers to the exercise above were, in fact, the logarithms-to-base-2 of the numbers in the exercise.

binary digit or 'bit'

Now if some letters are more probable than others it saves time, when sending a coded message, if the instruction-to-select devotes *less* time to the *more* probable ones than to the less probable ones. The Morse code is an example of this approach.

Table 2 shows the letters of the English alphabet arranged in order of probability. The high probabilities and most of the short Morse code are at the top of the table, and by and large the long codes (except notably O) are at the bottom. The dots and dashes of the Morse code words can be

Table 2

Symbol	Probability	Morse Code	Information per symbol
Space	0·187		2·46
E	0·107	·	3·22
T	0·086	–	3·84
A	0·067	· –	3·90
O	0·065	– – –	3·94
N	0·058	– ·	4·11
R	0·056	· – ·	4·16
I	0·052	· ·	4·27
S	0·050	· · ·	4·33
H	0·043	· · · ·	4·54
D	0·031	– · ·	5·02
L	0·028	· – · ·	5·17
F	0·024	· · – ·	5·38
C	0·023	– · – ·	5·45
M	0·021	– –	5·60
U	0·020	· · –	5·64
G	0·016	– – ·	5·94
Y	0·016	– · – –	5·95
P	0·016	· – – ·	5·95
W	0·016	· – –	6·02
B	0·012	– · · ·	6·42
V	0·007	· · · –	7·06
K	0·003	– · –	8·20
X	0·001	– · · –	9·54
J	0·001	· – – –	9·85
Q	0·001	– – · –	10·0
Z	0·001	– – · ·	10·5

Note: A probability of 0·187 is the same as a chance of 1 in about 5·3 since $1/0·187 = 5·3$. The information per symbol is then $\log_2 5·3 = 2·46$.

interpreted as instruction-for-selection and therefore represent rather crudely the information content per letter.

A more accurate representation of information content for each letter is given in Table 2. The figures are obtained simply by taking the log-to-base-2 of (1/probability), using more accurate values of probability:

$$I = \log_2(1/\text{probability}) \text{ bits/letter}$$
$$\text{or } I = -\log_2(\text{probability}) \text{ bits/letter.}$$

So, in engineering, the idea of information per digit has a precise meaning. *It is a simple function of the probability of the receiver inferring the digit correctly.* Thus the more the receiver knows about a structured language being used, the less the information per digit a message in that language contains.

The importance of all the foregoing is just this. Claude Shannon* in 1948 showed that, if you can identify the information content of a message, then in theory *it is possible to convey that information through any amount of noise and distortion simply by adding redundancy of the right kind to the signal.*

Somehow this happens in speech between humans. We can decipher speech through noise levels which would defeat the best machines we can devise. But then humans can bring so much knowledge to bear on the problem. All the rules of phonetics, syntax, semantics which each human being somehow 'knows', but cannot explain, are brought into use.

How, for example, do I know that 'I tried flying' means something quite different from 'I tried to fly', or that 'I like to run' means the same as 'I like running', or that 'I want to run' is all right whereas 'I want running' is nonsense?

Let us then accept that every human being 'knows' the secret of the use of redundancy in efficient signal transmission in noise, and look at technology's first hesitant steps towards doing the same kind of thing.

---

* Shannon, C. E. (1948). *Bell System Tech. J.* **27**, 379 and 623.

Section 7

Coding to correct or detect error

Engineers usually see to it that the information content per digit of a signal they construct is high. That is to say a signal consisting of five binary digits like 10110, as used in telegraphy (and transmitted electrically perhaps as voltage on, voltage off, voltage on, voltage on, voltage off) is concerned with an 'alphabet' of 32 letters or signs, including 'no sign', and it is assumed that each sign is as probable as any other one, so there is no redundancy. (If such a code were used with an alphabet of 16 digits, 16 of the possible code patterns would not be used.) Of course the letters have to be equiprobable from the receiver's point of view. If the receiver does not know English then he (or it) does not 'know' that E is more probable than Z.

So we will start our discussion of error-correcting and error-detecting codes by assuming that the message we want to send has been reduced to an initial efficient signal in which there is no redundancy whatever. We will assume a five bit binary code dealing with a 32-letter 'alphabet' and that there is no statistical structure in the message to be sent (i.e. no letters or letter sequences are any more probable than any other). Table 3 shows such a code.

We will consider three types of error-correcting or error-detecting code: the Parity Check Code, the Block Code and the Chain Code.

Table 3. *A telegraph code*

Letter	*Code*
A	11000
B	10011
C	01110
D	10010
E	10000
F	10110
G	10111
H	01101
I	01011
J	00101
K	01100
L	11010
M	11110
N	01001
O	00111
P	00110
Q	00011
R	11101
S	01010
T	00001
U	11100
V	01111
W	11001
X	10100
Y	10101
Z	10001
Carr. rtn.	00010
Line feed	01000
Letters	11111
Figures	11011
Space	00100

7.1 The Parity Check Code

The Parity Check Code simply involves appending a sixth binary digit to the 5-digit code to be transmitted *so chosen that the number of 1s in each code is an even number*. This extra digit is called *a parity check digit* and the task of the receiver of such a code is to ensure that there is always an even number of 1s in the code. If there is an odd number the receiver records an error, or transmits a message asking for a repeat transmission. This parity check digit is a redundant part of the signal from the point of view of the message because it is not always needed, but if an error occurs then it is needed to indicate the presence of this error.

parity check digit

Below are some examples of even-parity-check 6-bit codes:

Letter	*Code*	*Parity Check Code*
A	11000	110000
B	10011	100111
Z	10001	100010

There is an even number of 1s in each 6-bit 'word', so any single error either from 1 to 0 or 0 to 1 can be detected. Double errors cannot be detected, but triple ones can, etc.

7.2 Block codes

The parity check principle can be extended to a two-dimensional array. Suppose, as shown in Fig. 19, the 5-bit codes of a telegraphy signal are arranged in groups of five.

First, each 5-bit code has its parity check bit added.

A parity check bit is then added to achieve even parity for all the first digits (represented by the first column in the table), all the second digits, all the third digits and so on, as shown.

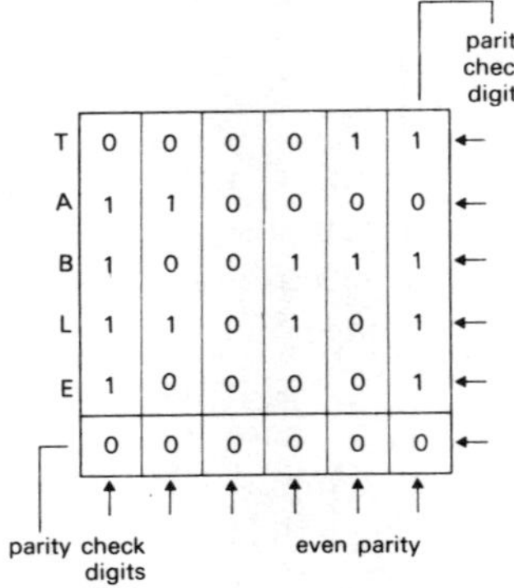

Figure 19   Block code. The word to be transmitted is TABLE. The non-redundant codes are therefore 00001, 11000, 10011, 11010 and 10000 (see Table 3). Eleven redundant digits are added.

Finally a parity check digit is added to ensure that the parity check digits themselves have even parity.

This is then an error-correcting system. If one error occurs anywhere in the block, it is possible to identify which element is in error and hence to correct it.

Does it follow that the bottom right-hand digit, giving even parity on the parity check digits, always agrees for both rows and columns?

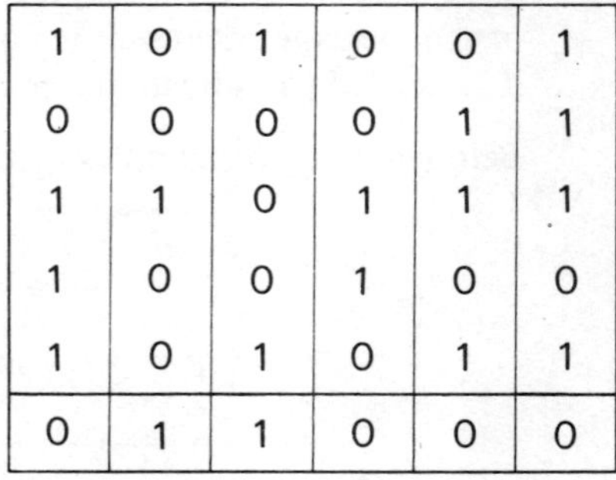

Yes, because either way this digit indicates whether there is an even or odd number of 1s in the message.

The diagram of Fig. 20 shows three block codes, in the first two of which there is only one error. Where has the error occurred? In the third block code something more disastrous has happened. Can you identify what has happened?

1	0	1	1	0	1
1	0	0	0	1	1
0	1	1	0	0	0
1	0	1	0	1	1
1	0	1	1	1	0
0	1	1	0	1	1

(a)

1	0	1	0	1	1
1	1	1	0	1	0
0	1	1	1	1	0
1	0	0	1	0	0
0	0	1	0	0	0
1	0	0	0	1	0

(b)

1	0	1	0	0	1
0	0	0	0	1	1
1	1	0	1	1	1
1	0	0	1	0	0
1	0	1	0	1	1
0	1	1	0	0	0

(c)

Figure 20   Block code exercises in error correction and detection.

In the first two blocks, it is simply a matter of searching for errors in the parity check digits or in the parity check digit applied to each of the parity check codes. In the first block there is an error in the parity check of the second row and the third column. In each case the number of 1s is odd instead of even. Thus there is an error at the intersection of the second row and the third column.

In the second block there is an error in the parity check digit in the fifth row and also in the column of parity check digits. The parity check digit for this column should be a 1. So the error is in the parity check digit of the fifth row, and there is no error in the message itself. Thus errors in the error-correcting digits are equally well detected.

In the third block there are errors in both the first and third rows and in the third and fifth columns. Thus there are *four* locations in which *two* errors might have occurred. So it is impossible to correct the error with certainty but at least the presence of two errors has been detected. (Note that there is nothing wrong in there being four 1s in the parity check column but only two 1s in the parity check row. It is only possible to ensure that there is either an even number or an odd number in the parity check row or column.) Block codes of this kind are called single error-correcting and double error-detecting codes. They are, of course, transmitted one row at a time.

7.3 Chain codes

A recently invented type of code which has found great use in space communication is the chain code. First we will consider how such a code is constructed and then look at some of its properties.

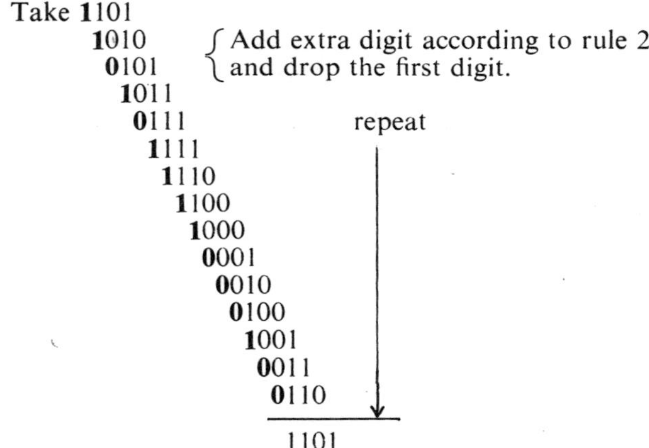

Figure 21   Generation of a 15-digit chain code for the message code 1101. The chain code is the sequence formed by the first digit of each row, namely 110101111000100.

The best way to explain the production of a chain code is by means of an example, so we will consider a 4-digit signal and proceed to construct the redundant digits belonging to this signal. Briefly what we do is this:

1 We start by writing down the initial 4-digit message to be transmitted, say 1101.

2 We add an extra digit to the end of this message according to the rule:

> If the first two digits are the same add a 0; if the first two digits are different add a 1. This rule is called modulo-2 addition – applied to the first two digits of the message.

Thus the message 1101 becomes 11010 because the first two digits are the same.

3 We take the last four digits of the new 5-digit signal and repeat 1, 2 and 3 until the original message is about to be regenerated.

These rules are quite clear cut so a computer, or simple electronic circuit can generate these extra digits.

Fig. 21 shows an example of this chain code generation process. The final code signal is the sequence of first digits of each line. In this case the coded signal is **1101011110000100** for the original message of 1101. The whole sequence is fifteen digits in length, so eleven redundant digits have been added to make the error correcting code.

> I would like you to try generating the coded signal for the message 1111. Compare it with the coded signal for 1101 and what do you notice?

The sequence you should have obtained is 111100010011010. There are several things you might notice. The most important for our purpose are:

(a) They are both the same sequence but each starts at a different place. (Draw the sequence in the form of a circle to see this clearly.)

(b) All but one possible sequence of 4-digits appears once, and only once, within the whole code sequence. So the same sequence is the coded signal for every 4-digit message code, except one.

> Which is the missing sequence. Can you see why it is not included?

As shown in Table 3 the message codes used in telegraphy are each of five digits. Chain codes for 5-digit messages can also be generated, but the rule is slightly different. It is:

> If the first and third digits are the same add a 0 (making six digits this time); if the first and third digits are different add a 1.

The first digit is then dropped and the process repeated with the remaining five digits.

SAQ 3

Generate the coded signal for the message code 11000 (i.e. the code for the letter A).

pseudo-random sequence

The missing sequence is four zeros. With this sequence the rule for generating a new digit always produces another zero.

For answers, see end of unit.

As before we can draw the sequence you should have obtained in a circle as shown in Fig. 22. The sequence is thirty-one digits in length and it contains within it every possible 5-digit sequence except 00000.

Now how is this used for error correction? We want to show that if errors occur we can say what should have been in their place. In fact the code will correct any seven errors – as explained below.

The property of these sequences which is used to prove the code's error-correcting capabilities is that if you take two such sequences and 'add' them together using modulo-2 addition again, then *another version of the same sequence will be produced*.

Modulo-2 addition, you will remember, is that if two digits are the same the sum is 0 and that if two digits are different the sum is 1. An example of this is shown in Fig. 23.

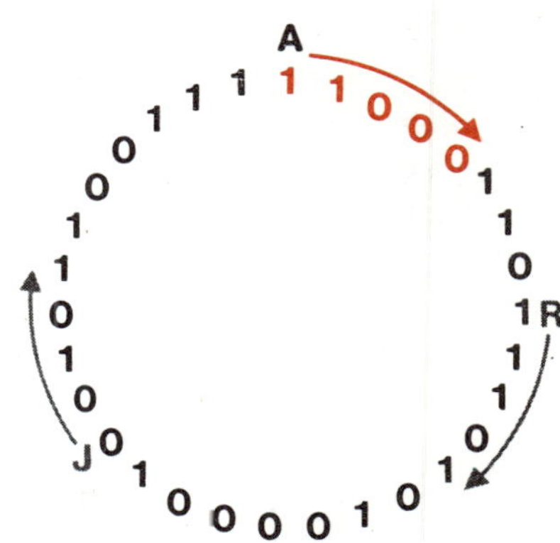

Figure 22   *The* 31 *digit chain code for 5 bit message codes. The sequence of digits is the same for all 5 bit codes, but starts at a different place. The starting points for A, R and J in telegraph code are shown.*

A = **11000**110111010100001001011001111
R = 11101010000100101100111**11000**110   This is A shifted 8 to the left

= J = 00101100111**11000**110111010100001   This is A shifted 11 to the right

Figure 23   *The upper code word is 'added' to the lower code word to produce a third code word. Each word is the same pattern but shifted to left or right. (The rules of 'addition' here are $0+0 = 0, 0+1 = 1, 1+0 = 1, 1+1 = 0$ but no carry.) That is, a 1 in the sum indicates a point of difference between the two sequences. If the two codes added are the codes for A and R as shown, the 'sum' is the code for J.*

The digits of the two codes are 'added' digit by digit, so that every 1 in the sum indicates a point of difference between the original two sequences.

What do you notice about the new sequence generated by this addition?

The sum is the same chain code shifted to left or right. This is a very remarkable property. There are always sixteen 1s (i.e. sixteen differences) between two sequences when they are 'added' in this way. *So every code sequence* (produced in this way) *differs from every other code sequence in sixteen places.*

Now after a code sequence has been transmitted it may contain errors, and so the job of the receiver is to decide which of the possible thirty-one signals was sent. The simplest rule to work with is to choose the code sequence *nearest* to the received one. Now since every code sequence contains sixteen different digits from all other ones it follows that for up to seven errors in any received pattern it is possible to decide correctly what the original code sequence was. For eight errors the code sequence is 'equidistant' from two sequences, and with nine or more errors the received signal is nearest to the wrong sequence. *Thus it is possible for a receiver to find the correct signal even if up to seven errors have occurred in transmission.*

How can the receiver do this?

The simplest way is to 'add' the signal (using modulo-2 addition) to each correct sequence. The sum with the fewest 1s indicates the correct sequence transmitted. This can very easily be done automatically – as described in Unit 13 later in the course.

Codes can be constructed for any length of message but they increase in length greatly as the message word-length increases and so does the number of errors they will correct. Indeed the length of a chain code is

2^n-1 digits, where n is the length of the message word. Thus a message of twenty digits produces a chain code of over 1 000 000 digits. Words of 8 digits – giving 255-digit sequences – are about the longest used in practice (in the Mariner space probe) and can correct up to sixty-three errors in the signal!

SAQ 4

The following fifteen digit chain code is received.

100011001100110

Find which 4-digit message was transmitted using this error correcting code.

SAQ 5

Find a formula for the number of errors a chain code of length 2^n-1 will correct.

SAQ 6

When you toss a coin the sequence of heads and tails which results is said to be 'random'. The sequence is characterized by the fact that whatever the result of previous coin tossings the probability of a head or of a tail remains 0·5 (i.e. a 50 : 50 chance) each time you toss the coin. Explain why chain codes are sometimes called pseudo-random sequences.

For answers, see end of unit.

In space communication the problem is to receive a message accurately despite the electronic 'noise' which emerges from space. It is always possible in principle to overcome noise by transmitting a signal with enough power to render the noise negligible. But where satellite transmissions are concerned, greater power is very expensive. (To double the available power in a spacecraft is likely to cost millions of pounds.) So an accurate method of transmitting signals in spite of the presence of noise is of great technical importance.

But the extreme resistance to error of these codes is just one aspect of their importance. It is, in a sense, just one manifestation of the fact that the message is spread out through the signal. It is possible to infer the 5 message digits correctly from anywhere in the 31 digit sequence provided that there are no more than 7 errors altogether and this is the basis of decoding it.

Section 8

Conclusion

We have described one of the most advanced and ingenious error-correcting codes used in communication at the present time. What has it achieved?

The first idea it illustrates is that it is possible to add error-correcting digits to a code pattern so that, if errors occur either in the pattern *or in the error-correcting digits*, the pattern can be reconstructed without error. An obvious technique like just repeating the pattern does not have this property because, even if only one error occurs, it is no longer possible to say which was the right original pattern.

The second idea is that when redundant digits of the right kind have been added, as in the chain code, the information in the signal appears to be spread out right through the signal. There is nowhere which can be called the redundant part of the signal and no parts which are clearly the information-carrying parts. Provided the point in time when the sequence begins is known (even if, in the case of the 31-digit code, it begins with seven false digits), the correct sequence can be determined and the signal decoded without error.

Speech signals are like this too, but more so. It is certainly true that we can usually 'decode' a sentence we have misheard provided that we have not misheard too much.

Indeed, since we receive phonemes at such a high rate during normal speech, it is probably true that we cannot hear all of them at any time and 'error correction' of what we hear goes on continuously. One of the clever things about speech, however, is that we do not have to be told when the signal starts and stops. In other words not only can we do error correction, but also we can deduce from the signal itself where it begins.

But speech is much cleverer than this. For it not only enables us to decode the words, but also tells us who the speaker is, what his mood is, and even when he does not mean what he says! And, it seems, it is to do with error-correction too. The speech signal can be very distorted by, say a telephone line, or very much affected by noise and yet we still seem able to obtain these kinds of information from it. So various differing types of information are all conveyed together in one signal, which itself is subject to very great detailed variability in the frequencies, amplitudes, timing, etc., of its components.

In other words, speech is a signal of the most extraordinary subtlety and complexity and yet is handled with consummate ease by all of us. There is clearly something here for technologists to study.

But the brain, which handles this signal, seems to have the same kind of properties itself. Although there are areas of the brain with functions which are quite specific, it seems to be true that significant areas of the brain can be put out of action without affecting many of its functions too much. Certainly individual cells can die at an alarming rate without the brain's functions deteriorating at all for many years. So it seems as though the many functions it performs are spread out, so to speak, over a large area of the brain in a complex interdependent way, just as information of various kinds is spread out in speech. However, the brain is still more subtle and complex because it can handle many kinds of signal – vision, touch, movement, etc., in much the same way. They are all signals with great variability even when the same information is being handled.

All this and more, as the first two units showed, is the concern of systems studies. We need a way of handling complex signals and systems which does not involve analysing them into many separate independent parts, since their very essence is their complex interdependence. This is what systems analysis is about, and what these first three units have been about.

Acknowledgement

I would like particularly to thank Bell Telephone Laboratories Incorporated for permission to base large sections of this unit on their excellent book, *The Speech Chain*, by P. Denes and E. Pinson. For those wishing to read more about speech production organization and recognition this book is strongly recommended. Paperback and hardcover editions of *The Speech Chain* can be purchased from Doubleday & Co. Inc., The Anchor Science Study Series, 277 Park Avenue, New York, New York 10017.

Answers to SAQs

SAQ 1 (Section 2.3)   The key in answering this question lies in doing a Fourier analysis of the vibrations exciting the pendulum: that is, finding out what harmonics the driving signal contains. A sequence of impulses, as in Fig. 12a, contains many harmonics as well as the fundamental frequency. A sinusoidal oscillation contains only the fundamental frequency.

If you could not answer the question before you read the preceding paragraph, pause now and have another try.

The answers to the four parts of the question are:

(a) Impulses at the resonant frequency of the pendulum 'contain' this frequency as the fundamental of the impulses' frequency, so this will set up resonant oscillation of the pendulum (but not at quite so great an amplitude even if the impulse amplitude is the same as a sinusoidal one – can you see why?)

(b) Impulses at one-half or one-third the pendulum's natural frequency have harmonics at the pendulum's natural frequency, so resonant response will occur. This is just like the stimulation of resonance in the vocal tract. The vocal cord pulses are at a much lower frequency than the resonances of the vocal tract.

(c) Impulses at higher than the pendulum's frequency do not contain 'subharmonics' (i.e. components *less* than the fundamental frequency) so the pendulum will not be excited.

(d) Sinusoidal oscillations contain no harmonics of any kind so sinusoidal vibrations of the pendulum support will *not* stimulate the pendulum into oscillation even though impulses will.

You can check all these statements with a simple home-made pendulum, except perhaps the last because a sinusoidal movement of the support is difficult to produce – you would need a wheel or piston drive, rather like that in Fig. 5.

SAQ 2 (Section 2.3)   Fig. 2 shows the amplitude of a freely vibrating system – such as a pendulum – whose natural frequency is 1 Hz. Fig. 2b shows a lightly damped vibration, Fig. 2c a more heavily damped one.

Fig. 3 shows the response of two resonant systems, both with a resonant frequency of 1 Hz, one lightly damped and therefore responding with a large amplitude to a small exciting vibration, and the other more heavily damped and unable because of the damping to respond with such large amplitude vibration.

Thus the two damped-responses of Fig. 2 might very well correspond to the two frequency-responses of Fig. 3, the curve of Fig. 2b corresponding to the sharp response curve (shown in black) of Fig. 3.

SAQ 3 (Section 7.3)   The first steps in the generation are:

```
11000
10001
00011
00110
01101
11011
10111
01110
```
                        and so on

Figure 24.

The complete answer is shown in Fig. 22.

SAQ 4 (Section 7.3)   In order to decode the received signal it is necessary to compare it with each possible correct sequence. The simplest way to do this with pencil and paper is to write the correct sequence down, twice, and in sequence. Then write the received signal on a separate piece of paper and compare it with each position of the double sequence. The position with fewest differences reveals the correct code. (A circular construction of the same process is more efficient but less easy to handle.) The following array illustrates the method and shows the position of the correct sequence

Correct sequences 1 1 1 1 0 0 0 1 0 0 1 1 0 1 0 1 1 1 1 0 0 0 1 0 0 1 1 0 1 0

received sequence (15 possible positions)   1 0 0 0 1 1 0 0 1 1 0 0 1 1 0

---

modulo-2 addition   0 0 1 1 0 0 0 0 1 0 0 0 0 0 0

Figure 25.

There are only three 1s in the modulo-2 addition shown revealing 3 differences between the correct sequence and the received signal. Any other placing of the received signal in relation to the correct sequences shows more differences. The 4-digit message that was to be transmitted is therefore 1011, the first four digits of the correct sequence.

SAQ 5 (Section 7.3)   First we must find the 'difference' between any two sequences. For the 31-digit code the difference was sixteen digits because 'adding' according to modulo-2 addition always gave a new version of the same sequence with sixteen 1s in it. The number of 1s in any code sequence is half of (one more than the number of digits in the sequence).

Therefore the number of 1s $= [(2^n - 1) + 1] \div 2$
$$= 2^n/2$$
$$= 2^{n-1}$$

(There is always an even number.)

The code will tell you correctly which original sequence an erroneous one is nearest, provided there are less than half the number of 1s in the sequence (i.e. less than half the difference between any sequence and its neighbours). In the 31-digit code there are sixteen 1s, so seven errors could be corrected.

Number of correctable errors $= [(\text{number of 1s})/2] - 1$
$$= (2^{n-1}/2) - 1$$
$$= 2^{n-2} - 1$$

e.g. if $n = 8$, then sixty-three errors can be corrected (255-digit code). (If you are not familiar with 'indices' like n in 2^n refer back to Modelling I and then return to this SAQ. The important property to grasp is that $x^n \div x = x^{n-1}$.)

SAQ 6 (Section 7.3)   If you study either of the chain code sequences we have considered and regard them as the results of coin tossing experiments (e.g. 1 = head, 0 = tail) you might easily conclude that the sequence was random. That is:

(a)   The number of 1s is approximately equal to the number of 0s.

(b)   The number of times a 1 follows a 0 is approximately equal to the number of times a 0 follows a 0. Similarly, the number of times a 1 follows a 1 equals the number of times a 0 follows a 1.

(c)   Select any pattern of 1s and 0s (e.g. two 0s, two 1s, 01, 110, etc.). The next digit is as likely to be a 1 as it is a 0.

Thus this property, which characterizes random sequences, also characterizes chain codes even though chain codes are not random at all. Chain codes are therefore sometimes called pseudo-random (binary) sequences.

ACKNOWLEDGEMENTS

Grateful acknowledgement is made to the following sources for material in this unit:

ILLUSTRATIONS

Bell Telephone Laboratories Incorporated for Figs. 2, 3, 8a, 8b, 9, 10, 13, 14, 15, 17a and 17b; Penguin Books Ltd. for Figs. 4, 5 and 6

T100—THE MAN-MADE WORLD

Technology Foundation Course Units

Week number	Correspondence text		Unit number
1	Systems		1
2	The human component		2
3	Speech, communication and coding		3
4	Statistics	(first part of)	10
5	Production systems modelling		28
6	The production environment		29
7	Systems File		5
8	Mechanics		6
9	Electricity and magnetism		7
10			8
11	Energy conversion		20
12	Power and society		21
13	Environment File		23
14	Maintaining the environment		26
15			27
16	Noise abatement		26/27S
17	Economics File		11
18	Structures and microstructures		9
19	Materials		22
20	Chemical reactions		24
21	Chemical processes in industry		25
22	Cities File		30
23	Automatic computing		12
24	The heart of computers		13
25	Computer systems		14
26	Economics of traffic congestion		31
27	Transport File		31 File
28	Reliability	(second part of)	10
29	Analogue computing		15
30			
31	Control		16
32	Modelling II		18
33	Design		32
34			33
			34

as you opened to the sun, and they have dropped into the earth. Your body is the earth, your baby the seed. New life is arising!

16. *Be a holy person.* Now become a holy person, sitting alone but not feeling alone, for you are filled with the realisation that Life fills every cell of your body, and through it you connect with all other living things. There is deep peace in your being. Act out within yourself how it might feel to be such a person. How do they feel within themselves? What expression is on their face? How do they sit? How do they arrange their hands as a posture of how they feel inside? Let it happen to *you*!

Some people find that when they act out these postures and movements, tears of sorrow or joy are released. Or else spontaneous movements begin to occur to them. In a few cases, it is as if they were merely a spectator while Life in them moves their body.

If this happens to you, it is quite safe to allow it to happen, although it may seem strange. But it is no stranger or miraculous than the movements Life is continuously performing with your body. Breathing, eating, digestion, heartbeat, are all movements Life makes spontaneously. Being used to these, we do not feel them as strange. We *expect* them. During birth our body also goes into spontaneous contractions, and again we *expect* this. But inasmuch as we can trustingly go along with such spontaneous movements, to that degree, Life and you can co-operate to give birth to your baby.

The only difficulty occasionally arising is that these spontaneous movements lead to things we might not judge to be good. When we eat something poisonous, Life in us causes us to retch and vomit. This is an unpleasant feeling and experience, but it is Life's way of healing us and protecting us. During our life we may have 'swallowed' many poisonous emotions, desires, thoughts or experiences. When we relax and let Life act on us or flow through us, as in these postures, it sometimes tries to have us 'vomit' up these past poisons. Thus if we lost someone we love, and buried our pain and grief, we may, if we let spontaneous movements or emotions